NASHVILLE

Joel + Debbie,

My long time friends,

thanks for all your support.

Bob Saly 9.16.90

Amy Grant

Margt Nil

"Partners in Progress"
by Walter Jowers

Windsor Publications, Inc.
Chatsworth, California

NASHVILLE

UPBEAT AND DOWN TO BUSINESS

A CONTEMPORARY PORTRAIT

BY AMY LYNCH AND MARGARET E. DICK

Windsor Publications, Inc.—Book Division
Managing Editor: KAREN STORY
Design Director: ALEXANDER D'ANCA
Photo Director: SUSAN L. WELLS
Executive Editor: PAMELA SCHROEDER

Staff for *Nashville: Upbeat and Down To Business*
Manuscript Editors: DOREEN NAKAKIHARA AND SUSAN M. PAHLE
Photo Editor: ROBIN STERLING
Senior Editor, Corporate Profiles: JUDITH L. HUNTER
Production Editor, Corporate Profiles: ALBERT POLITO
Proofreader: MARY JO SCHARF
Customer Service Manager: PHYLLIS FELDMAN-SCHROEDER
Editorial Assistants: KIM KIEVMAN, MICHAEL NUGWYNNE, MICHELE OAKLEY, KATHY B. PEYSER,
THERESA J. SOLIS
Publisher's Representatives, Corporate Profiles: JOHN COMPTON, JACK FOWLER,
WILLIAM METCALFE, AND JOHN ROSI
Layout Artist, Corporate Profiles: CHRIS MURRAY
Layout Artist, Editorial: MICHAEL BURG
Designer: ELLEN IFRAH

Windsor Publications, Inc.
ELLIOT MARTIN, Chairman of the Board
JAMES L. FISH III, Chief Operating Officer
MICHELE SYLVESTRO, Vice President/Sales-Marketing
MAC BUHLER, Vice President/Acquisitions

Library of Congress Cataloging-in-Publication Data
Lynch, Amy.
Nashville : upbeat and down to business / by Amy Lynch and Margaret E. Dick ;
    "Partners in progress" by Walter Jowers. —1st ed.
p.152 cm.23x31
Includes bibliographical references.
ISBN 0-89781-360-X
1. Nashville (Tenn.)—
    Pictorial works. I. Title.
HC108.N2L96   1990   90-32323
330.9768'55053—dc20   CIP

ISBN: 0-89781-375-8

RIGHT: *Dawn breaks over the Cumberland River, bathing the region with a gold and radiating light. Photo by Matt Bradley*

PAGE SIX: *The yards and porches of Nashville's homes are places of community pride, marking the growth of families and the events that shape their lives. Photo by Bob Schatz*

PAGE NINE: *Deaderick Street in the heart of downtown Nashville is lined with stately glass-adorned towers. Photo by Bob Schatz*

PART ONE OPENER: *Early morning mist settles over the rolling hills of this Brentwood farm. Photo by Bob Schatz*

PART TWO OPENER: *Modern office towers rise into the Nashville sky, blending the designs of the city's contemporary architecture with the styles of the past—a city rich with a sound business community and a spirited economic future. Photo by Bob Schatz*

This book is dedicated to Phil and Steve,
who knew us when . . .

Contents

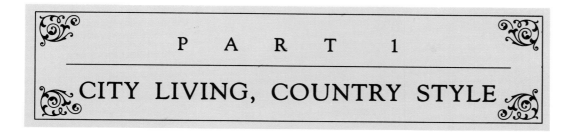

PART 1

CITY LIVING, COUNTRY STYLE

PART 2
NASHVILLE'S ENTERPRISES

PART

1

CITY LIVING,

COUNTRY STYLE

A Community Emerges

T he making of a city is not easy, but Nashville's citizens have been equal to the task. From the beginning, they met adversity with energy, creativity, and tenacity. The result is a thriving, lively city, tempered by its trials and nearly bursting with promise. Lots of adjectives apply to Nashville's rough-and-ready past—turbulent,

raucous, even glorious. But not boring—never boring.

INTO THE VALLEY OF THE CUMBERLAND
The pace of life in Middle Tennessee was slower in 1779 than it is today, but things weren't any easier. Nashville's first permanent settlers arrived that year, a development that did not please the Chickasaw, the Shawnee, or the Creeks—all of whom had claimed and had hunted the area for centuries. Smallpox, Indian attack, hunger, and cold claimed 40 of those first settlers before they could even finish their 1,000-mile journey along the Tennessee and Cumberland rivers to the French Lick, as they called the Nashville site in those days. Other pioneering sorts came overland through the Cumberland Gap and met with less mishap, though tradition dictates that they trudged through snow to cross the frozen Cumberland River on Christmas Eve.

No sooner had the settlers begun to build their outposts, than the Indians began to harass them. For years the various Indian groups in the area had accepted white hunters without much

hostility, but women, children, cabins, and palisades meant permanence. Encouraged by both the Spanish and the British, the tribes fought with a vengeance.

The settlers' unguarded crops and livestock were destroyed. Then people working in the fields were killed. Indian ambush took a steady toll. No one was safe outside the outposts, and sieges brought death inside as well. Food and gunpowder ran low. A few at a time, French Lick families began to flee to safer ground in Kentucky and Illinois.. The stalwarts who stayed banded together in Fort Nashborough, within blocks of today's downtown. Considering that Indians in the area outnumbered settlers by better than a hundred to one, the wonder is that the settlement survived at all.

But in 1784, when North Carolina deeded 640 acres to each settler still in the French Lick region, there were Englishmen, Scots, a few Germans and Welshmen, and one Frenchman there to claim the fertile, rolling hills. By then the British had lost their hold on the area; the Indians were less hostile; and the community changed its name

---

Thousands of spectators flocked to witness the dedication of the Andrew Jackson statue on the capitol grounds in 1880. Courtesy, Tennessee State Library and Archives

A visit to the Hermitage, home of Andrew Jackson, provides an engaging glimpse into Nashville's past. Photo by Bob Schatz

to Nashville, dropping the "borough" as a result of anti-British sentiment.

Growth and prosperity followed. The city was officially laid out in 200 one-acre lots along the bluff, with four acres reserved for a public square. The first merchants arrived. Soon there were grain mills and whiskey stills. One of the first settlers became the first sheriff. A doctor came, and then another.

As settlers competed for land, title disputes were inevitable. Lawyers who migrated to Middle Tennessee found plenty of work. A court was set up to settle civil and criminal suits. The first school was chartered and endowed with 240 acres of land. Churches and public buildings were erected. Stores sold dry goods. By 1786, as many as 800 settlers lived in the area. Nashville was gradually losing its rudest frontier trappings and taking on the appearance of a backcountry village.

By 1800 Nashville had a post office, a newspaper, several stores and taverns, and an inn. There were a few touches of luxury in the muddy streets: a coach or two, some carriages, even men wearing silk stockings and brass-buckle shoes. In another decade the town had its first bookstore. By that time the area boasted 9,000 residents, and Tennessee had become the 16th state of the Union. Nashville's day was dawning.

STEAMING ALONG: THE AGE OF JACKSON

By 1800 keelboats and flatboats took cotton, tobacco, and a variety of goods downriver from Nashville to New Orleans. Some boats were poled up the Cumberland, bringing supplies from the Ohio River Valley. But nothing floating on the river had ever caused as much excitement as did a very special arrival on March 13, 1819. People crowded the banks as around the bend, big wheel churning and whistle blowing, a steamboat laden with goods chugged its way to a berth at the foot of Broad Street. For all practical purposes, that day marked the end of frontier isolation.

That maiden steam voyage from New Orleans to Nashville was just the beginning. In no time, steamboats reduced the distance between the growing settlement and the rest of the country. City wharves were soon crowded with steamers, and enormous warehouses sprang up along the riverfront. Nashville was the undisputed commercial center of the region.

Nothing could have been more appropriate than that particular boat's arrival in March 1819. The man for whom the steamboat was named was a Nashvillian already famous the country over and celebrated in Europe as well. It may have been the boat's name, as much as anything else, that drew spectators to the Nashville dock for its arrival. It was the *General Jackson*.

By 1820 Andrew Jackson's popularity had given Nashville considerable national prestige. A brash, redheaded lawyer and public prosecutor, Jackson cut a formidable figure in the new frontier. After coming to national attention during the Creek Indian War in Alabama, Jackson roundly trounced the British in New Orleans at the end of the War of 1812. He was wildly embraced as a national hero.

In an emerging nation still totally dependent on military power for survival, Jackson was a driven and devoted general. The people loved him. Winning his early races easily, Jackson served successfully as a senator and congressman. When he indicated that he was ready for the presidency, his constituents and his party were ready to support his bid.

Jackson's star rose no more dramati-

cally, however, than that of the town he called home. The mid-1800s brought unprecedented development for Nashville. Increasing traffic on the Cumberland lent the city new importance and prosperity. Wealthy businessmen built lavish estates in town, while rich farmland supported grand plantations in the surrounding countryside. A medical school was founded. Theaters offered Shakespearean plays. A board of education was established, and P.T. Barnum brought Jenny Lind to town.

James Robertson (1742-1818)

Sometimes called "the father of Tennessee," James Robertson was not a flamboyant man, nor was he well educated. Yet his talents were exactly what the fledgling settlement of Nashborough needed. Robertson was a supremely talented leader, effective and indefatigable.

It was Robertson who led early settlers overland to the Cumberland Valley in 1779 without losing a single member of his party. Two weeks later he helped draft the Cumberland Compact, Nashville's earliest statement of government. According to tradition, the stalwart Robertson was one of 20 men cut off from Fort Nashborough by the Chickasaw during what is called the Battle of the Bluffs. Robertson's wife is said to have loosed the fort's dogs on the Indians, distracting them long enough to allow her husband and his companions to reach safety inside the fort. When Indian attacks frightened away most of the original settlers, Robertson stayed. He established a relationship with the Chickasaw leader, Piomingo, and negotiated peace with that warring tribe.

Robertson also negotiated on the settlers' behalf with an even more formidable foe—the government of North Carolina, which claimed the French Lick region. With his guidance, early Nashvillians laid legal claim to their hard-won land. Robertson never stopped pioneering. In 1818, while serving as government envoy to the Chickasaw,

Courtesy, Tennessee State Library and Archives

he died on the bluffs above the Mississippi River where present-day Memphis stands. It is safe to say that without James Robertson there would not have been a Nashville—or a Tennessee—as we know it.

The Nashville Public Square was a beehive of activity in the early days of the city's development. Courtesy, Tennessee State Library and Archives

Most importantly, Nashville was chosen as the state's capital city in 1843, and the skyline was soon dominated by a Greek Revival capitol building erected on the city's highest hill.

A final footnote, and one of the most significant to that era, was the coming of rail traffic. In December 1850 a steamboat called at the Nashville wharf to deliver a locomotive engine. That steamboat, in effect, brought to town the very instrument of its own destruction. By the turn of the century, Nashville would be a rail center rather than a river town. In 1851 a train made its first trip, an 11-mile run to the neighboring settlement of Antioch. Three years later a railroad line was opened to Chattanooga. A depot was built, and the first passenger trains chugged to its platform. In 1859 a line was opened between Louisville and Nashville, connecting Nashville with cities to the

north. It was finished just in time for the Union army to take it over. The Civil War had begun.

A CITY OCCUPIED

In 1860 Nashville was not eager for war. There was general support for the practice of slavery in the region, but not a willingness to secede from the Union or to go to war for it. In spite of Middle Tennessee's rich farms and plantations, slave owners were a minority in the region's white population. Furthermore the 17,000 citizens of Nashville were too prosperous to have a compelling reason to go to war. Nashvillian John Bell ran for president in 1860 as the candidate of the Constitutional Union Party. Bell's stance for preservation of the Union above all else won him the state's electoral votes.

The winner of that election, however, was a lawyer from Illinois by the name of Lincoln. And when he called for Tennessee troops to put down the insurrection in South Carolina, Nashvillians balked. Along with the rest of Tennessee they passed a resolution to leave the Union, the last state to do so.

Water and rail access made Nashville the state's strategic manufacturing and stockpiling center. Southern troops trained in the streets, a local powder plant was converted to a munitions factory, and women made uniforms. There were huge parades and demonstrations. When Confederate general Albert Sidney Johnston marched into town to take charge of military forces in the area, he was cheered by thousands.

But Nashville's enthusiastic rebellion lasted only eight months. The city was poorly defended. In February 1862 federal gunboats swept past Fort Donelson downriver and steamed to the vulnerable city on the bluff. Johnston pulled out, and total panic set in. A mass exodus began as military units scattered in disarray. In less than a week, Nashville was transformed from a wartime boomtown and Confederate arsenal to a ghost town.

Then Union troops marched into the city, and Tennessee's capital became a federal stronghold in the Southern interior. And so it remained for the duration of the war.

Nashville's three years of occupation changed the city forever. A gracious, settled town became a military center—frenzied, rough, and muddy. Public buildings housed Union soldiers. Churches and hotels became army hospitals, and the city's population swelled to an unmanageable 80,000. The air and water were dirty. Disease ran rampant. Forts were thrown up on the hills, and the capitol was fortified with cannon.

The federal army cut down the forest of trees that had surrounded the city and lined its streets, making wasteland of the countryside so that dug-in Union troops could see threatening rebel forces coming. There was good reason for such precaution. Confederate cavalry regularly raided the outskirts of town, and in 1864 Confederate troops under General John B. Hood tried unsuccessfully to reclaim the city in one of the bloodiest battles of the war.

It has been said of Nashville that her head was with the Union, but her heart was with the South. The city was, in

Enormous warehouses sprang up along the banks of the Cumberland River in Nashville as the growing city evolved into a regional commercial center in the early 1800s. Although the advent of the railroad would soon become the leading force in local trade, the Nashville Wharf, pictured here around the turn of the century, continued as a major player in the city's economy. Courtesy, Tennessee State Library and Archives

William Strickland (1788-1859)

When the state legislature decided in 1843 that Nashville was to be the state's permanent capital, a national search was begun for an architect to design a capitol building. William Strickland of Philadelphia, once an apprentice to the designer of the nation's Capitol, was chosen for the job. In 1845 he moved to Middle Tennessee to supervise construction. Strickland designed other Nashville landmarks, but the capitol was to be his masterwork.

Built of limestone, the beautifully proportioned Greek Revival statehouse holds its own today. In a downtown full of skyscrapers and postmodern, broken pediments, the graceful lines of the capitol's classic facades show to good effect. By his own request, Strickland's vault lies in one of the capitol walls. No other architect has ever had such a profound effect on the city skyline. During the mid-1800s, as Nashville sought an identity, Strickland gave her a face.

Courtesy, Tennessee State Library and Archives

fact, at war with itself. Sarah Childress Polk, wife of the former president, received Confederate and Union generals alike at her Nashville home, but she was an exception. Neutrality was not a luxury afforded the average citizen. Military governor Andrew Johnson required loyalty oaths of all officials, educators, journalists, and clergymen. A secret police force dealt with those who resisted. Everywhere families were divided as fathers, brothers, and cousins took up arms on opposite sides of the conflict. The war inflicted scars that would not heal for generations.

Yet when peace finally came, Nashville was physically intact. The city had escaped the devastation suffered by other Southern cities. Because the town had surrendered virtually without resistance, its citizens had been treated relatively well. Most Nashvillians had been able to stay in their homes, and many merchants had continued to do business. They even had prospered. That strong economic base proved a boon to the town's reconstruction as, for the next 10 years, Nashvillians picked up the pieces.

CENTENNIAL CITY

After having been occupied in an unfinished state by soldiers and prisoners during the war, the Maxwell House Hotel opened in 1869 to great fanfare. A symbol of the city's determination to rise again, the Maxwell House established a standard of excellence and was the city's premier hotel for nearly a century. Eight American presidents stayed there. The hotel served a special blend of coffee that President Teddy Roosevelt was to declare, "good to the last drop."

When Nashville turned 100 in 1879 it was the South's fourth-largest city, with 40,000 people. Nashvillians marked their centennial with a month-long celebration. Thousands crowded downtown streets to see parades and to visit a large exhibition hall on Broad Street. A striking equestrian statue of Andrew Jackson—a symbol of unity, vitality, and better times—was erected on the capitol grounds.

Signs of growth were everywhere. The city had emerged from the Civil War with a diverse economy, no longer agrarian. The local printing industry was expanding. Two major universities, Vanderbilt and Fisk, were established shortly after the war. Financial institutions began to prosper. Doctors and public officials worked to make the city cleaner and healthier. Easy water and rail transport encouraged even more wholesale and distribution activity.

By 1889 the city's first electric streetcars, decked with ribbons and flags, carried passengers along major arteries. Ten years later, a huge passenger depot called Union Station was opened downtown by the Louisville and Nashville Railroad (L&N). The lavish, Romanesque Revival station featured stained-glass skylights 63 feet above the lobby floor, digital clocks in a 220-foot tower, and gold-leaf detailing on its muraled walls. Most Nashvillians had never seen anything like it.

They hadn't seen the likes of the battle that was shaping up over control of the city's rail traffic either. Because of its central location and role as a distribution center, Nashville's history has always been written in part by the means of transportation available to it. The city's turn-of-the-century economy depended heavily on railroad companies like the L&N and Nashville, Chattanooga and St. Louis Railroad (NC&StL). Controlled by out-of-town interests, those companies came under fire from Nashville businessmen who started their own competing rail company, the Tennessee Central. The battle

was bitter on both sides, but in the end, the massive L&N maintained control over Nashville's lucrative rail traffic.

In an effort to gain the goodwill of Nashville citizens, L&N officials helped organize and fund the state's centennial celebration of 1897. One hundred years after Tennessee became a state, Nashvillians built a special park, lakes and fountains, exhibit halls, and a midway to celebrate the occasion. The centerpiece of the celebration was a full-scale replica of the Parthenon. Festivities lasted for six months and attracted 1.7 million people. The extravaganza

The Legislative Lounge in the state capitol echoes with the voices of days gone by. Photo by Matt Bradley

cost more than a million dollars, and actually turned a modest profit.

When other exhibit halls were dismantled the following year, the Parthenon was left standing. Nashville liked it. In the early 1920s the city decided to replace the temporary wooden structure built for the centennial with a permanent concrete building, encouraging the use of a nickname the city also liked—the Athens of the South.

MUSIC CITY U.S.A.
AND THE TWENTIETH CENTURY

More than 15,000 Nashville-area men and women participated in World War I, and while they were away the city's population swelled with an influx of factory workers, many of them employed at the huge Dupont munitions plant built in a bend of the Cumberland River. Nashville's prosperous German community found itself beleaguered by anti-German sentiment during those years, and several institutions got new names as a result. The German Methodist Church became Barth Memorial, and Edward Potter's German American Bank became Farmers and Merchants.

Other controversies rocked Nashville during the early years of this century.

In celebration of the state's centennial in 1897, the residents of Nashville built a special park in which a full-scale replica of the Parthenon was erected. The original wooden structure was replaced with a permanent concrete building in the early 1920s, and today stands as a monument to the city's glorious past. Photo by Matt Bradley

Jim Crow segregation was enacted, the temperance issue caused bitter political divisions, and suffragettes marched in the streets. The times were clearly changing.

An enterprising dealer drove the town's first Ford up the steps of the capitol, and soon cars were commonplace. Even airplanes didn't arouse much comment. Nashville's diversified economy, burgeoning insurance

corporations, and cluster of banks and investment companies lifted the state capital to the top rank of American financial centers, earning the city yet another nickname—the Wall Street of the South.

The Great Depression hurt Nashville, but recovery was relatively swift. One reason was that the town's insurance industry had remained extremely strong in spite of the national recession.

Oddly enough, it was music that kept it that way.

In 1926 radio station WSM began broadcasting the Grand Ole Opry, a program of fiddle and string-band "barn dance" music. The National Life and Accident Insurance Company based in Nashville owned WSM. People across the country responded to WSM and, as a result, to National Life. When company salesmen went calling

Modern office towers accentuate the expanding Nashville skyline. Photo by Bob Schatz

21

Shown here shortly after its opening around the turn of the century, the majestic Union Station, built by the Louisville and Nashville Railroad, was designed in the Romanesque Revival style. Courtesy, Tennessee State Library and Archives

Parachutes, airplanes, naval vessels, combat boots, and sandbags poured out of the city. As products left, soldiers streamed in.

MIDDLE TENNESSEE MECCA

On December 7, 1941, news of Pearl Harbor reached town, and the first troop train passed through Union Station that night. Within months, three large military installations were located near Nashville. Some 600,000 men took part in area war games, often invading citizens' lawns, fields, and barns. Locals dubbed it "the second Yankee invasion," while military newspapers called Nashville a "soldier's mecca."

The city was flooded with uniforms. As many as 80,000 servicemen left bivouac and visited Nashville each week—quite a crowd in a city of only 167,000. Not only were soldiers trained or stationed in Nashville—thousands more traveled through on furlough, on their way to duty, or on their way home. Most of them stopped at Union Station, which was often so mobbed that people could hardly move. Soldiers and civilians lined up by the hundreds at ticket windows, open day and night. Railroad employees worked around the clock while weary soldiers waiting for trains slept wherever they could find space. Red Cross workers regularly took hundreds of loaves of bread, whole cases of eggs, and hundreds of pies to the train station to feed hungry soldiers stranded there. When peace came, Nashvillians surged into the streets and formed impromptu parades. At Union Station, train whistles and bells rang all night long.

THE SUNBELT IN MODERN TIMES— THE METRO EMERGES

Those whistles blew in a city ripe with opportunity, but fraught with problems. Technology was literally taking off. The first jets soon landed at Nashville's newly expanded airport. Television came to town in 1950, and

in the hills of Pennsylvania or on the plains of Illinois, WSM listeners bought policies. Thus country music, as the genre came to be called, helped Nashville out of the Depression and gave the city yet another nickname— one destined to stick. Today people who have never heard of the Wall Street of the South or the Athens of the South respond immediately to Music City U.S.A.

By 1940 the newspapers reported that the business volume of local manufacturers, wholesalers, and retailers had exceeded pre-Depression levels for the second consecutive year. Bank clearings surpassed the billion-dollar mark for the first time in city history. Nashville had not only recovered, it was positioned to prosper. World War II enhanced that position. As the war began, local manufacturers turned their efforts to military production.

Robert Penn Warren (1905-1989)

The Southern Renaissance, a flurry of literary activity centered at Vanderbilt University during the 1920s and 1930s, produced a host of tremendous writers. But the most famous is probably Robert Penn Warren.

Born in the countryside just north of Nashville, Warren enrolled at Vanderbilt University in 1921. He intended to study chemical engineering, a notion that lasted, according to his own reckoning, about three weeks. Warren's ambitions changed because he encountered two teachers, John Crowe Ransom and Donald Davidson, the nucleus of a group of poets called the Fugitives.

Warren was still writing poetry four years later when the national press descended on tiny Dayton, Tennessee, to cover the Scopes Trial (dubbed the "monkey trial" by reporters). While lawyers debated the teaching of evolution in the public schools, stories in national newspapers portrayed Tennesseans as quaint, ignorant, and lazy—an image that offended Warren and his scholarly friends.

Warren, Ransom, Davidson, and nine other Southerners spent the next five years developing a strong and complicated response to that image. Their book was called *I'll Take My Stand*, and they certainly did. The Agrarians, as this group came to be called, argued that Southern society fostered individual integrity, stability, and aesthetic qualities alien to the industrialized North. The book produced as heated a controversy as any Southern publication printed before or since.

In spite of criticism, the Agrarians continued their scholarly defense of Southern traditions and values. Warren continued writing poetry and began producing novels as well. He won the Pulitzer Prize for fiction in 1947 for *All the King's Men*, and in 1958 *Promises*, a volume of poetry, won him both a Pulitzer and the National Book Award. In 1979, just before he celebrated his 74th birthday, Warren was awarded a third Pulitzer for another book of poems, *Now and Then*.

Along the way, he has taught and inspired a new generation of Southern writers—Randall Jarrell, James Dickey, and Peter Taylor among them. With fellow Fugitives and Agrarians, Warren ushered in a new era of Southern literature and, in the process, put Nashville on the literary map.

---

the first skyscraper went up in 1957. In the 1960s three major interstates came together downtown, leaving Nashville at a crossroads—in more ways than one.

Labor unrest, racial inequality, crime, and corrupt ward politics haunted urban Nashville. Soldiers newly returned from war were acutely aware of the city's deficiencies in public health, education, and housing. The city threatened to become dangerous and unlivable. A vast slum lay in the very shadow of the capitol, yet city officials lacked funds for urban renewal. Black and white children attended schools that were separate and anything but equal. Public transportation and government services were inadequate to serve the city's population. In short, Nashville faced the problems that challenged nearly every other urban center in the country during that period. Something had to be done.

The census of 1960 told it all. For the first time in history, Nashville's population declined. More people lived in surrounding Davidson County than in the city proper. Those workers and their families who had come to Nashville during wartime had, for the most part, peopled the suburbs and beyond, shunning the decaying inner city.

Fortunately, a new generation of city leaders was working on solutions. Their first step was to extend the city limits to the county line. City and county governments became one—a streamlined system dubbed "Metro." The change didn't come easily. Consolida-

tion met with great resistance from city and county alike. Yet once it came, the reformation did away with ward politics, effectively ending the reign of corrupt local "bosses." Under the new system, wasteful city and county duplication and competition were eliminated, and a more efficient government made new money available for improving city services.

Urban renewal proved the second solution to Nashville's inner-city crisis. Federal and local programs eliminated the worst slum areas, cleaned up

downtown, and provided adequate public housing for the poor. A strong historic preservation movement worked hand-in-hand with urban renewal to make the city livable again. Beginning in the early 1970s, Nashville's oldest neighborhoods underwent transformation as young professionals began to move in and restore old homes. Downtown's oldest buildings, especially the warehouses along the waterfront, were reclaimed as well.

The third solution to Nashville's postwar crisis came with the civil rights movement. Beginning in the early 1960s, blacks staged sit-ins and marches to protest their status as second-class citizens. Early protests were nonviolent, but as the struggle grew more intense, rioting occurred and the National Guard was called in to keep the peace. By the early 1970s, after a decade of conflict, Nashville's schools were finally desegregated and its restaurants served black people as well as white. No longer were washrooms or water fountains marked "whites only." Racial segregation was ended as a matter of law, though the dilemma of race endures.

All of these changes brought about a more unified, equitable, and livable city, and just in time. Hard on the heels of the social reform of the 1960s came growth—tremendous growth—throughout the Sunbelt during the 1970s and 1980s. The U.S. population

shifted southward during those years, and Nashville fairly exploded. The skyline grew taller, and green space diminished. Metro's population neared half a million. Contiguous counties added half a million more.

Today dozens of international enterprises are headquartered in Middle Tennessee, and high-tech industry is commonplace. Automotive companies, most notably Nissan USA and GM's Saturn Division, have settled in the area. In 1987 Nashville completed a new airport. International flights leave its runways.

Nearly incomprehensible change has taken place in just 200 years. Today's *General Jackson* is a showboat for tourists, and today's Fort Nashborough is a museum. But some things remain the same. The region's central location and fertile, rolling countryside made Nashville a bustling hub of trade and culture from the beginning, and these same factors currently shape its future. And if those early days were marked by trial, they were transformed by vitality. Even during the worst of times, Nashville has never stopped growing. Such vitality bodes well for the city's third century.

This young boy enjoys a quiet moment with a treat of fresh popcorn in downtown Nashville. Photo by Matt Bradley

Thomas G. Ryman (1841-1904)

Young riverman Tom Ryman came to Nashville during the Civil War to take advantage of the demand for fish among the military population. Though his family was penniless, Ryman worked hard and shrewdly. By 1865 he had bought his first steamboat and began hauling cargoes along the Cumberland. Beginning in 1875 he organized three packet lines, consolidating them in 1885 as the powerful Ryman Line. That same year, Captain Tom found religion and closed down the saloons on his 35 steamboats.

Ryman's mansion stood atop Nashville's Rutledge Hill so that he could watch his paddlewheelers come and go at the wharf below. From there he could also see another project dear to his heart—the construction of an elaborate temperance hall on Broad Street. Ryman called it the Union Gospel Tabernacle. His dream was that evangelists would hold revivals there.

When Ryman died 15 years later, mourners packed the hall, and the minister who had converted Ryman suggested that the hall be renamed after its builder. The tabernacle became Ryman Auditorium.

Over the years, use of the building broadened. Theodore Roosevelt, Enrico Caruso, and Sarah Bernhardt drew crowds there. But the Ryman achieved its real fame from performers of a different nature—performers with names like Grandpa Jones and Hank Williams. What had begun as fire and brimstone ended with fiddle and ballad. The Grand Ole Opry broadcast from the Ryman for more than 30 years. Today Ryman's hilltop mansion is gone, but his auditorium is a musical landmark.

25

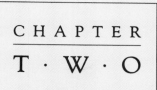
Downtown's Spirited Revival

Broadway. It's nine blocks of Nashville's past, present, and future. When you come into the downtown on Broadway you pass the beautifully renovated Union Station; Hume-Fogg, a high school for gifted students; and the refurbished Customs House. As you crest the hill at Seventh Avenue, you can see all the way to

the Cumberland River. On the left is the new Nashville Convention Center and Stouffer Hotel complex. On the right, closer to the river, sits the elegant Merchants Restaurant. At the foot of the street is Riverfront Park.

The Italianate and Victorian buildings on Broadway date back to the time when Nashville was a bustling river port. Hotels and warehouses were built close to the river to serve the constant flow of steamboats. The same area was later known as "Furniture Row" because all of Nashville's major furniture retailers had stores there.

At night, music moved the street. Ryman Auditorium, the former home of the Grand Ole Opry, sits right around the corner on Fifth Avenue. Popular singers regularly dropped into nearby Tootsie's Orchid Lounge between Opry shows to entertain the crowd there. Across the street, in Ernest Tubb's Record Shop, a live radio show drew fans well past midnight.

But Broadway began to change in the

late 1960s, when several of the furniture stores closed their doors and followed their customers to the suburbs. The final blow to the street came in 1974 when the Opry moved to its new home north of town. The fans followed. Tootsie's has survived as a tourist stop, but Ernest Tubb moved his radio show to a new location near the Opry. Broadway became home to bars, pawn shops, and empty storefronts.

But even as Broadway was fading, active plans were being made to rejuvenate it. A downtown plan adopted in 1977 proposed attractive mixed-use development for the street. The city agreed to commit public funds to enhance the area for private investment.

Since 1983 the city has spent more than one million dollars on streetscape and facade improvements along Broadway. In return, private developers have invested almost four million dollars restoring the old buildings for new uses. The undesirable businesses are being replaced by upscale restau-

---

The Shelby Street Bridge offers this breathtaking view of Nashville's downtown along the Cumberland River. Photo by Matt Bradley

rants and shops.

The transformation and rejuvenation of Broadway is one example of the many public-private partnerships that have revitalized downtown Nashville. The new Nashville Convention Center and Stouffer Hotel complex, Union Station, the redevelopment of Second Avenue, Riverfront Park, and Church Street Centre are projects that demonstrate the public and private sectors's shared commitment to Nashville's continued progress.

Downtown Nashville has been planning for its future since 1949, when its first long-range plan was approved. Throughout the 1950s, 1960s, and early 1970s, Nashville made judicious use of federal urban renewal money to undertake improvement projects downtown. In a massive project that began in 1951 and took six years to

LEFT: A favorite Nashville landmark, the beautifully renovated Union Station on Broadway reopened as a hotel in the mid-1980s. Photo by Bob Schatz

OPPOSITE: Nashville's modern skyscrapers, historic warehouses, and landscaped parks all help to illustrate the dynamic past and promising future of this progressive southern city. Photo by Bob Schatz

BELOW: Businesses are returning to downtown Nashville, bringing with them a welcome surge in employment and new office construction. Photo by Rudy Sanders

The Man With The Plan

When Richard Fulton, former mayor of Nashville, was growing up in East Nashville he used to ride the streetcar into town. "Actually, I hung on the back. If the street conductor didn't see me I'd get a free ride into the city. Otherwise it cost a nickel."

Fulton says he loved to come downtown on Saturday afternoons to go to the movies. "Everyone would be downtown doing their weekly shopping. It was very exciting just to be there," he remembers. "On Sunday my friends and I would walk all the way to Union Station just to watch the trains come and go."

During his 12 years as mayor, Richard Fulton

Photo by Bob Schatz

Photo by Bob Schatz

set about to re-create the city he enjoyed so much as a boy. It wasn't an easy job. When Fulton first took office he decided to convert Church Street, the cramped and congested main route into town, into a landscaped outdoor mall. Decorative lights and planters appeared. The asphalt was torn up and a special bricklayer hired to brick the street. Nearly everyone thought the redevelopment of Church Street was a terrible idea. Some called it "Fulton's Folly." But Fulton persevered.

Now Church Street is home to the $30-million Church Street Centre, which is reviving the downtown retail market. One major office tower has been completed on the street and two more are planned. Fulton's Folly indeed.

As a former congressman, Fulton brought to the mayor's office a broad perspective of what would work in a downtown and what wouldn't. He was willing to take dramatic steps to get people interested in the downtown. He set an agenda for downtown Nashville and he saw it through. The Nashville Convention Center complex, Riverfront Park, and the redevelopment of Second Avenue, Broadway, and Rutledge Hill all were spearheaded by Fulton.

"The core of a city is like the core of an apple. If the core of an apple becomes rotten the entire apple will soon be rotten. A city is only as healthy as its core. If the downtown suffers it's only a matter of time before the suburbs will suffer too."

The Dominion Bank building is a striking addition to the Nashville skyline. Photo by Matt Bradley

complete, the slums and dilapidated buildings behind the state capitol were replaced with landscaped slopes, high-rise buildings, and a tree-lined boulevard. The Capitol Hill Redevelopment was the first urban renewal project approved in the United States.

The widening of Deaderick Street, the construction of Legislative Plaza in front of the capitol building, and the upgrading of the public square near the Metro Courthouse were all funded with federal dollars.

Despite these efforts, in the late 1960s the downtown hit hard times as the suburbs began to boom. The first suburban mall opened, and Saturday shopping in the downtown became a thing of the past. Many retailers followed their customers to the malls, leaving empty stores behind.

The downtown employment base of finance, insurance, and government workers remained strong, but office parks began luring many other businesses to the suburbs. To add to the downtown's

RIGHT: Built primarily in the 1870s and 1880s, the structures of historic Second Avenue are rich with architectural detail. Photo by Bob Schatz

BELOW: Once a warehouse district, the historic buildings along Second Avenue have been converted into attractive shops, offices, and apartments. Photo by Matt Bradley

woes, the federal urban renewal program was eliminated.

When Congressman Richard H. Fulton became mayor in 1975, he made the downtown his number one priority. He immediately commissioned a downtown study. "I wanted projects that would capture the public's imagination and make them want to come downtown again," he said. Fulton also wanted to persuade the private sector that the downtown was a viable alternative to the suburbs.

Development proposals included in the 1977 Downtown Nashville Plan and Development Program brought a new emphasis on using public expenditures to leverage private investments. This plan detailed a series of projects, to be completed over 15 years, which would enhance the economic vitality of the downtown area. The projects included the redevelopment of Church Street, a riverfront park, the adaptive use of Second Avenue and Broadway as mixed-use developments, the realignment and widening of Commerce Street, a pedestrian link from Church Street to the riverfront, a convention center, and a parking garage.

Under Fulton's leadership more than 90 percent of the proposed projects were completed in less than 10 years. The city committed more than $120 million in public funds to these improvements; in turn, private developers spent in excess of $500 million on downtown projects.

Perhaps no other project captured the hearts of Nashvillians the way Riverfront Park did. Fulton wanted to give people a reason to come downtown again. With Riverfront Park he succeeded.

For years a number of proposals for riverfront development, including restaurants and residences, had been explored and rejected. Instead the city decided to develop the 6.5 acres of riverbank as a park—a place where nearby office workers could get away from the hustle and

Urban Pioneers

"When Bob and I first moved downtown my parents wouldn't tell anyone where we lived," remembers Lisa Schatz. "Now," she adds with a laugh, "they brag about us!"

Lisa and Bob Schatz moved downtown in the late 1970s. They were among a handful of urban pioneers who made their homes in converted warehouses along Second Avenue.

After living in several apartments along the street, Lisa and Bob bought their own building, the former Standard Sales Co. warehouse. On the ground floor they opened the Market Street Emporium, with space for small specialty shops and restaurants. Bob, a photographer, has his studio on the second floor.

Lisa likens downtown living to living in a resort town. "There's always something to do. It's quite festive. We have a lot of fun here." She adds, "I'm a recreational shopper so it's a great place for me. I used to drive to a mall. Now I just hop on the trolley."

In the evenings Bob and Lisa, along with their toddler Douglas and dog Max, enjoy walking around the downtown and visiting with friends. "I grew up in the suburbs. The only time I saw my neighbors was when they'd drive by and wave from their car. Downtown we always see people we know and we always stop to chat. There's a real community down here," says Lisa.

That community includes everyone from unmarried young professionals to families with children, as well as a retired couple or two. In past years the demand for downtown housing has boomed. More than 400 new units—both condominiums and apartments—have been added.

Most of the new construction is concentrated along Second Avenue toward the Rutledge Hill area. Many people would like to see additional

Photo by G.W. Austin

housing units in the heart of the downtown, but current zoning doesn't permit that use. Judy Steele, assistant director of the Urban Development Office, hopes that proposed revisions in the zoning codes will allow for housing. "If we're going to be a 24-hour city we simply must have more places throughout the downtown for people to live."

More people living downtown means that important suburban services like grocery stores and theaters will move downtown. That's a prospect that delights Lisa Schatz. "I'll never have to leave the downtown. Everything I need will be right here."

Riverfront Park has been a catalyst for renewed interest in the downtown area ever since it was completed in the summer of 1983. Photo by Bob Schatz

bustle of the city and relax.

On opening day for the park, July 10, 1983, a full schedule of activities was planned. The river was full of pleasure boats. Clowns and jugglers entertained in the streets; a fireworks display was scheduled for the evening. "We imagined that maybe 10,000 people might come downtown to see the new park," remembers Mary Anne Harwell, who was a city consultant at the time. "Over

40,000 people showed up!"

For many it was their first trip downtown in years. People were surprised at the progress. Families strolled along Second Avenue visiting the shops and restaurants located in the former warehouses. Many walked to Legislative Plaza and went wading in the fountains there.

Cecil Herrell, director of development for Nashville's Urban Development Office, sees Riverfront Park as

Completed in 1859, the Tennessee State Capitol is one of the oldest functioning antebellum capitols in the country today. Photo by Bob Schatz

the catalyst for renewed interest in the downtown. "I think people were generally pessimistic about the downtown until Riverfront Park," he said. "It gave us the momentum we needed."

The park received an award from the U.S. Department of Housing and Urban Development in recognition of its public-private partnership achievement. Herrell estimates that the city's investment of $5 million in Riverfront Park stimulated more than $80 million in private investment in the adjoining Second Avenue and Broadway historic districts.

Nashvillians soon learned to live with the sights and sounds of construction throughout the downtown. During a five-year boom in the mid-1980s, three office towers were constructed adding 1.4 million square feet of new office space. The convention center and adjoining 700-room Stouffer Hotel opened for business. Several blocks of warehouse space on Second Avenue were adapted for use as offices, shops, and restaurants. Seven new apartment and condominium projects, with over 400 units, were constructed. Union Station, Nashville's favorite landmark, reopened as a hotel.

Businesses like Aetna Insurance and IBM, which had departed the downtown for the green grass of suburban office parks, returned. Downtown employment grew from 45,000 to over 55,000. It's expected to hit 65,000 by the end of this decade.

As Nashville prepared to enter the 1990s, good things continued to happen in the downtown. Early in 1989 Church Street Centre, a 180,000-square-foot downtown shopping mall, opened for business. Located on Church Street at the foot of Capitol Boulevard, Church Street Centre is an easy walk from major employment centers. It's linked to the downtown's only remaining department store, Castner-Knott, by a skywalk. A second glass-enclosed skywalk connects the mall to the Stouffer's Hotel and convention center complex. The mall features more than 70 specialty

The Cookie Man Can

Christie Hauck remembers quite well the very first customer who walked into his downtown cookie store. "I was standing in my brand new store feeling a little anxious about everything. This lady walks in, looks around, and asks how much a cookie costs. I told her about 55 cents. She looked at me and laughed. She said no one in Nashville would pay that much for one cookie. Then she walked out of my store."

Christie grabbed a tray of cookies, stood outside, and started to give cookies away. At least he tried to give them away. He had the dubious misfortune of opening his cookie store on the very day that nearly everyone in Nashville started on the Rotation Diet, with announced plans to lose a collective one million pounds in three weeks. "People would see me with the cookie tray and cross the street. They didn't want to be tempted," remembers Hauck with a laugh.

Once the diet was over, most people returned to their old eating habits and Christie Cookies was an instant hit.

Although he originally planned to locate his first store in a suburban mall, Christie now credits his early success to his downtown location. "There were 40,000 or 50,000 people walking around down here looking for something fun to eat. It was a tremendous untapped market."

From their comfortable offices in the Nashville Incubation Center downtown, Christie Hauck and his small staff oversee the company's retail, wholesale, corporate, and mail order sales. In 1985, its first year in business, the company

Photo by Bob Schatz

sold around 90,000 cookies for $50,000 in sales. Four years later the company sold 2.2 million cookies and recorded $1.5 million in sales!

A self-confessed "cookieholic," the remarkably slender Hauck says he downs three or four cookies each day "for quality control." He touts downtown Nashville to anyone who will listen. "It's the place to be. It's the center of activity for Nashville."

shops, a large fast-food court, and restaurants. Several retailers who had followed their customers to the suburbs have returned to the Church Street Centre.

Perhaps the most telling sign of the downtown's success is Opryland USA's interest in pursuing a redevelopment project here. The project has begun with a one-million dollar renovation of the Ryman Auditorium. It's anticipated that the project, which will include three office towers, specialty retailing, entertainment facilities, and possibly housing, will take several years to complete and will cost about $200 to $300 million.

Mayor Bill Boner has appointed a task force to study the possible expansion of the downtown convention center. Opened only three years ago, the facility has enjoyed enormous success. The expansion may include a sports arena. Mayor Boner, an avid basketball fan, says he'd like to attract an NBA expansion team to the city. "If we can get the arena I think we have an excellent chance at getting a team," says Boner.

As the downtown takes on a new look it remains very much a people oriented city. The wide, attractively landscaped sidewalks invite walking. It's a compact city. A walk from Broadway to the capitol building takes only a few minutes. Christie Hauck, who operates several gourmet cookie stores in the downtown, observes "It's a great place just to run into people. Everyone walks around— lawyers, bankers, everyone. Anyone I need to talk to I can meet on the street."

Mayor Boner says that the success of the downtown is a good sign for the suburbs. "All the truly great cities in this country have strong, vibrant, and exciting downtowns. In the last 15 years we've completed some wonderful projects in the downtown, which have really contributed to the quality of life here."

The vivid imagery of this street mural adds a splash of color to Nashville's downtown. Photo by Matt Bradley

Music—
And So Much More

Nashville's popular image as the home of country music can be deceptive. Country music certainly thrives in Music City, adding its down-home texture to the cultural fabric, but country is primarily a tourist attraction, an industry, and an export item. Natives regard the genre with tolerant affection, but they may not be fans.

You can find live, foot-stompin' performances any night of the week if you look for them, but there's a catch. Once you start looking, you're bound to discover other things as well—jazz, Shakespeare, ballet, rock, modern dance, and Bach. The Grand Ole Opry packs 'em in from around the world, but the locals, as often as not, enjoy the opera too. These days, there's plenty of variety for tourist and native alike.

For more than a decade, variety and growth have been the catchwords of Nashville's cultural scene. The Sunbelt boom of the 1970s took in Music City, and Middle Tennessee emerged as a business center of considerable influence. Against that backdrop, Nashville's arts community grew by leaps and bounds. New performing groups emerged. Others turned professional and became more prolific. Galleries mushroomed. High-quality touring productions began to include Nashville on their itineraries.

Television tapings and video production took off. As the city expanded, so did its audiences for the arts.

A STAGE IS BORN
At the heart of Nashville's cultural community is the Tennessee Performing Arts Center (TPAC), giving heart to those who dance and sing and declaim upon the stage. In just a decade, TPAC (pronounced "tee-pack") has virtually transformed and certainly enlivened the city's performing arts community. During the 1988-89 season, more than 600,000 people attended some 300 events in the center's theaters and concert hall. TPAC's busy, world-class stages have offered new opportunities to Nashville's opera associations, professional ballet and theater companies, city symphony, and local rock and country groups as well as traveling Broadway productions and international dance and theater troupes. From hoedown to highbrow, they all play at TPAC. During the early 1970s Nash-

Classical, traditional, and contemporary works are performed each season by the talented Nashville Ballet Company.
Photo by Bob Schatz

ABOVE: A flourishing theater community in Nashville includes the Circle Players, pictured here in the popular production Bus Stop. Photo by Bob Schatz

RIGHT: The Nashville Symphony has evolved into a professional full-time organization, now led by renowned conductor Kenneth Schermerhorn. Photo by Matt Bradley

ville developers Robert Hart and Victor Johnson were men with a plan. They wanted to give land to the city for a performing arts complex.

But Johnson and Hart had trouble finding anybody to buy into their dream. The problem was money. A first-class facility costs tens of millions to build, and arts centers are traditionally deficit-prone. The chamber of commerce didn't think it could raise the money, and the city didn't have it either.

Johnson invited the wives of a group of prominent local businessmen to a luncheon meeting, and he told them about the situation. The "city fathers," he explained, had turned him down. Longtime patron of the arts Martha Ingram was in the audience, and it was she who responded first. "How about the 'city mothers'?" she asked. "If we can raise the money, will you make the same offer to us?"

Essentially, Ingram bypassed the city

and went to the state. She asked Governor Winfield Dunn to designate state bicentennial celebration funds for the proposed arts center. That way, reasoned Ingram, the center could serve Tennesseans, not just Nashvillians. The governor liked the idea, but he couldn't earmark funds for a center that wasn't part of the network of state government buildings.

Ingram had an inspiration. "Why not combine the center with a proposed state office building, one slated to be built within a few years?" she asked. Before long, a deal was struck. The office building would be constructed as a tower above the performing arts center, which would be at ground level. And below ground, Tennessee would build a new state history museum, a nice complement to the performing arts center.

In the fall of 1980, TPAC opened its doors. That fragile first season began with a concert by the Cincinnati Pops Orchestra. "That was a thrilling day," remembers Ingram, "and it's thrilling still." For a decade now, TPAC has experienced slow but steady growth, its audiences and itinerary expanding with each new season.

THE PERFORMING ARTS: A SEASON OF GROWTH

Just 10 years ago, the Nashville Symphony was a part-time amateur group. Today it is full-time and professional. Growth that rapid can't always go smoothly. In 1980 the symphony's annual budget was only one million dollars, but that amount had grown to four million by 1988. Suddenly everything came to a halt. It became apparent that the symphony had over-extended itself financially. Threatened by a potential $700,000 budget deficit, the group suspended its 1987-88 season.

Within weeks, however, things got back on track. Today, running on an entirely feasible budget of three million dollars and offering a full season of classical, pop, and cabaret performances, the Nashville Symphony once again brings music to the city.

Much of the success of the Nashville Symphony is due to its conductor, Kenneth Schermerhorn. Schermerhorn was music director of the American Ballet Theatre for 11 years. He conducted the Milwaukee Symphony for 12 and is credited for its development into one of America's major orchestras. That's exactly what local

Light and shadow illuminate these graceful ballet dancers on the landscaped grounds of Cheekwood. Photo by Bob Schatz

41

Country music performer Roy Acuff embodies the heart and spirit of the Grand Ole Opry. Photo by Bob Schatz

folks hope he'll do in Nashville, where he spends most of his time rehearsing, conducting, and playing a leading role in the city's artistic development.

Orchestral offerings over the years have been varied and interesting, as the conductor has sought to build a strong city symphony with deep-rooted local support. And, far from objecting to country music, Schermerhorn often includes it in symphony programming. That's exactly the kind of versatility that has made Kenneth Schermerhorn internationally famous.

Slowly but surely, the dynamic conductor has guided the ensemble toward becoming a full-time professional organization as opposed to the part-time, amateur orchestra he found here—or as he puts it, he has been "turning a

An extravaganza of carnival rides, games, food, and non-stop entertainment, the colorful Italian Street Fair is held along First Avenue at Riverfront Park. Photo by Bob Schatz

"Minnie Pearl": Sarah Cannon

People still laugh every time Minnie Pearl barges onstage with her trademark bellow of "Howdee!" (decided accent on the *dee*). The country gal from Grinder's Switch, famous for the $1.98 price tag swinging from her straw hat, has been a regular on the Grand Ole Opry since 1940, and though she doesn't sing (not seriously, anyway) she is for many fans the very essence of country music.

The wisecracking, man-hunting Minnie Pearl probably hasn't told a new joke in years. She doesn't need to. Her material, always cornball and always kind, is old but funny still. Like Minnie's full-skirted gingham dress, they are country to the core.

Minnie's real-life incarnation, Sarah Cannon, however, is not.

In many ways the antithesis of her stage character, Cannon is happily married. She lives next door to the Tennessee governor's mansion in a gracious home with a tennis court and a swimming pool. A daily reader of the *New York Times*, and an articulate speaker, she is active in city charity circles in spite of her busy performance schedule.

In many ways, Sarah Cannon personifies two Nashvilles—the country music mecca that fans and performers come here to find, and the

Photo by Bob Schatz

midsize, mid-South city that most natives experience. One thing is for sure, on both sides of the fence, they all love Minnie . . . and Cannon, too.

nighttime orchestra into a daytime orchestra."

During Schermerhorn's six seasons here, the Nashville Symphony has seen its share of difficulties, but it has seen tremendous growth as well. In the process, music—country and otherwise—has benefited from his efforts.

All across Nashville, the proliferation and growth of performing arts groups during the past decade have been equally dramatic. Launched in 1986, the Nashville Ballet staged *Cinderella*, its first full-length production in 1988. The ash girl won her prince, and the company collected good reviews and revenue from

sold-out houses. Each season's calendar now includes ambitious classical and contemporary works as well as full-scale traditional presentations by the 18-member troupe.

Since 1984, Tennessee Dance Theatre has combined seminal works from the history of modern dance with themes and stories taken straight from the American South. And their appeal is more than regional. The *New York Times* and *Dance News*, among others, have praised this individualistic, tenacious, and growing troupe.

Opera, too, has seen a season of enormous growth. Since 1981 the Nashville

From Minnie Pearl's Museum and Gift Shop to the studio where Elvis recorded Are You Lonesome Tonight?, Nashville's Music Row offers a wide range of attractions for the avid country music fan. Photo by Bob Schatz

Opera Association has presented full-scale productions in TPAC's concert hall. And newer on the scene is Tennessee Opera Theater, a group emphasizing lighter operatic works done in English.

On the dramatic side of things, the Tennessee Repertory Theater consistently sells out its productions of proven favorites such as *Evita*, *To Kill a Mockingbird*, and *You Can't Take It With You*. More than 60,000 people view company productions each year.

Community theater flourishes on TPAC stages as the Circle Players, active since the 1940s, entertain with perennial favorites. The Nashville Academy Theater, founded in 1936, is one of the oldest children's theaters in the country; alternative theater finds expression at Actor's Playhouse, which lends itself to more intimate, offbeat, and improvisational productions. There's even Theatre Parthenos, Greek theater on the steps of the Parthenon.

MUSIC ON THE THIRD COAST
Symphonic classics may seem worlds away from barn-dance fiddlin', but in Nashville the two thrive side by side. Country music coalesced as a musical form during the 1920s, when Nashville radio station WSM gave banjo pickers and mountain-style singers a chance to strut their stuff. Today most of the country music played on radios around the world is recorded on Nashville's Music Row (in the general vicinity of Sixteenth and Seventeenth avenues), but the pervasive presence of country doesn't hurt the symphony a bit. As a matter of fact, it helps.

TPAC performance space is often rented to country music groups for television or radio tapings, and Chet Atkins has written original music for the Tennessee Dance Theatre. The music industry draws top-notch musicians to the city, and as often as not, the string player gettin' down in a recording session has had classical training. Conversely, the quartet member presenting Mozart on Sunday afternoon in the park may jam in a club on Saturday night. It makes for a rich mix.

Besides recording studios, Music Row houses music publishers, licensing agents, and museums by the dozen. There's RCA's Studio B, where Elvis recorded "Are You Lonesome Tonight?" and where Eddy Arnold, Jim Reeves, Dolly Parton, and Chet Atkins gave the genre landmark hits. Just down the

Prominent in gospel music for seven decades, the Fairfield Four Gospel Quartet gained national fame in the 1940s and has maintained a popular and long-lived success. Formed in 1920 at Nashville's Fairfield Baptist Church, the membership of the group today includes six members. Photo by Bob Schatz

street is the Country Music Hall of Fame and Museum, a center for the scholarly study and documentation of country from its bluesy-folksy roots to its current superstar status. Visitors shouldn't miss the gold cigarette lighter plucked from the wreckage of Patsy Cline's downed plane, and they can't miss Elvis' gold-plated Cadillac.

Country, of course, isn't the only music industry in town, it's just the biggest—and its spillover into other areas is considerable. Plenty of blues, folk, and gospel groups work out of Nashville, including the Fairfield Four Gospel Quartet.

In a town bristling with musical groups, the Fairfield Four stands out from the crowd. When the gospel quartet sang at the 1987 New Orleans Jazz and Heritage Festival, the crowd gave them a standing ovation and kept them on stage past their slotted time, yelling "One more, one more." The response was the same when the group sang at Carnegie Hall during the Smithsonian Folk Festival.

In fact, the Fairfield Four is as much in demand today as they were during their heyday in the 1940s, when they won a Colonial Coffee talent show,

leapt to national fame, contracted to sing on radio station WLAC, and toured coast to coast giving rhythmic, earthy performances. In 1942 they recorded "Don't Let Nobody Turn You Around" for the Library of Congress and, 50 years later, the group's fiery version remains an audience favorite. The secret of such long-lived success, according to the group's senior member, the Reverend Samuel McCrary, is working "by the spirit."

These days, McCrary's spirit-filled, emotive delivery is generally considered an American musical treasure. Prominent in gospel music for seven decades, the group formed in 1920 at Nashville's Fairfield Baptist Church. Members have come and gone (at one point the quartet even took a 20-year break), and most members are in their 60s or 70s. Today the quartet has five members. Says James Hill, who joined in the 1940s, "If the crowd enjoys it, we don't get tired. Shoot, we'd sing all night if they let us."

Rock music is also part of Nashville. More than one rock band has gotten its start on Music Row. There's even an annual Nashville Rock Extravaganza, which draws agents from Los Angeles and

New York to sign rock bands on the rise. Producers from both coasts regularly record acts in friendlier, less expensive Music City, earning Nashville a new nickname for the 1990s—the Third Coast.

Another high-tech, country music spin-off is the Nashville Network, a cable television station featuring country and seen in more than 40 million homes. The network is part of Opryland USA, a multimillion-dollar spillover of the music industry and another institution that has left its impression on Nashville's cultural milieu. The theme park at Opryland

presents not just country music, but Broadway-style shows like 1989's *Music! Music! Music!*, seen by more than half a million people.

THE VISUAL ARTS

For more than a century, Nashville has called itself the Athens of the South. But the local version of the Parthenon still comes as an anachronistic surprise. A full-size re-creation of Athena's shrine isn't exactly what one expects to encounter in a mid-South city. But there it sits, its friezes bustling with Greek deities, its beautifully proportioned doric columns mirrored in Centennial Park's reflecting pool.

Constructed as the arts building for the state's 1897 Centennial Exposition, the Parthenon has been in Nashville so long that it has become part of the city's identity, part of the local mythology. Whole generations of Tennesseans first discovered the arresting beauty of classical architecture on school field trips to Centennial Park. Half a lifetime ago they sat on the grass beneath the cool shadow of the Parthenon, ate their bologna sandwiches, and felt the power of this strangely placed icon.

For Nashvillians, the Parthenon has become a sort of cultural touchstone, a symbol of the city itself. It is on one of the porches of the Parthenon that Robert Altman's 1975 film *Nashville* reaches its tongue-in-cheek conclusion. It is here that more than 200,000 visitors come each year, here that Nashvillians bring their out-of-town friends who want to see the city.

The Parthenon stands unsurpassed as a powerful symbol of Nashville's arts community, but it is joined by other significant works and collections throughout the city. Within the Parthenon itself is the Cowan Collection, a group of outstanding American images from the late nineteenth and early twentieth centuries. Regional works are there too, as well as a collection of pre-Columbian artifacts from western Mexico.

The works at Cheekwood's Tennes-

The focal point of Centennial Park, the magnificent Parthenon stands as a monument to classical architecture and serves as the city's foremost art museum. Photo by Matt Bradley

see Botanical Gardens and Fine Arts Center comprise the city's most impressive collection and represent America's art history from Gilbert Stuart to Andy Warhol. Cheekwood also possesses one of the finest decorative arts collections in the country. Finally, the Stieglitz Collection—101 works by American Modernists and Europeans such as Cézanne, Renoir, and Picasso—hangs in the Van Vechten gallery at Fisk University.

The end result is a pleasant rounding out of Nashville's visual arts collections. American art is well represented—from its beginnings (in the Cheekwood Collection) to its adolescence (in the Cowan Collection) and its more modern manifestations (at Van Vechten). And then there are the galleries.

The most avant-garde of the visual arts are on display in Nashville, too. The state of the arts in Nashville changed dramatically a decade ago when numerous galleries burst onto the city's burgeoning arts scene. The initial pioneers in this area, Cumberland Gallery and Studio L'Atelier, have been joined by other fine galleries including Ambiance Gallery, Zimmerman-Saturn, Diops, and the Metro Arts Commission

Gallery. Each supports and provides a unique forum for contemporary artists. Adventuresome exhibits, works on paper, and postmodern constructions share space with conventional genres as the galleries create a growing visual arts market in Middle Tennessee.

Museums play a significant role in Nashville's visual arts community. The Tennessee State Museum is one of the finest state museums in the country. With its impressive collections of silver and historical portraits, the museum—while primarily a history museum—stands on its own as an arts center, too. The Cumberland Science Museum (complete with planetarium) is a mecca where schoolchildren explore the wonders of the natural world. Add to that the Museum of Tobacco Art and History, the Country Music Hall of Fame and Museum, and a host of smaller institutions, and you begin to get an idea of the diversity and richness of Nashville's visual arts community.

SUMMER LIGHTS: BRINGING IT ALL TOGETHER

The amazing thing is this: Diverse as it is, Nashville's arts community manages to come together, to present itself as a

The handsome plaster replicas seen at the Nashville Parthenon are direct casts of the original sculptures that adorned the pediments of the Parthenon in Athens, Greece. Photo by Bob Schatz

The Cheekwood Fine Arts Center houses a permanent collection of American art from the nineteenth and twentieth centuries, with an emphasis on the work of the region's leading artists. Photo by Bob Schatz

High-Impact Whimsy: Artist Lanie Gannon

Lanie Gannon specializes in the unpredictable. First, there's her unlikely combination of fine art with carpentry. Everyday materials—paint, springs, hinges, and wood—make up her playful, high-impact sculptures. The effect is dramatic.

Given to understatement, Gannon admits, "My work hits you right away." Her beautifully crafted, primary-colored pieces are tauntingly vivacious, conveying an energetic spirit within a slickly finished shell.

Then there's her humor. Gannon's radio series, for example, includes "Harpo Radio," "Oh Radio," and "Free Radio." But while her pieces provoke a smile, they demand deeper examination. Working with impersonal, ubiquitous objects, Gannon interprets them in very personal ways. The result is a satirical but nonthreatening comment on modern culture.

During the past few years Gannon's elusive, double-edged whimsy has earned her national attention, but her beginnings are in Tennessee, where she graduated from the Memphis College of Art and the Joe L. Evins Center for Crafts. Nashville's Cumberland Gallery was the first gallery to represent her. All in all, she has found Nashville a good place to work.

"The arts community here is very close," Gannon explains, "and it's growing. More and more artists are choosing to live and work here rather than taking off for bigger cities. There's no telling what may come of that. Just 50 years ago

Photo by Bob Schatz

country music had a very small beginning, too, but it has had international impact. The same thing could happen here. In fact, though you might not expect so, the musicians in town make it easier for the visual artists to work here. We are all of us working on the strength of a vision. We encourage each other."

more or less unified whole, once each year. The occasion is the annual Summer Lights Festival, staged by the Metro Nashville Arts Commission.

Think of Summer Lights as a sort of Nashville Spoleto, an elaborately choreographed festival of the considerable talent this city has to offer. The event absolutely takes over downtown streets for three rather intense days each June. More than 200 musical acts—including rock, pop, country, classical, and jazz—take to outdoor stages. The

visual arts are displayed inside and out. And everywhere, dancers dance, actors act, and improv groups have a heyday. Promising young musicians get their first big breaks, and established stars play to the crowd.

It's quite a crowd to play to. Ten thousand people showed up for the first Summer Lights in 1981, but 600,000 came in 1988. Like the rest of Nashville's arts community, Summer Lights has seen a decade of dramatic growth.

Anne Brown, executive director of

the Metro Nashville Arts Commission and the driving force behind Summer Lights, calls the festival a way of "deformalizing and demystifying" the arts—an idea that neatly complements the increased accessibility to the arts in general and that has transformed Nashville's arts scene during the past 10 years.

Brown wants even more to happen. The commission plans to extend the festival's run and to introduce a Winter Lights celebration as well. In fact, Brown sees the festival's roaring success as a sign that Nashville is ready to support the cultural and arts district currently developing downtown. She would also like to see an arts plaza and museum space established in the heart of the city. If the pace of the past few years is any indication, she—and Nashville—may not have long to wait.

LEFT: The annual Summer Lights Festival is a three-day event that celebrates the community's visual and performing arts. Photo by Bob Schatz

BELOW: Zimmerman-Saturn on historic Second Avenue is just one of the many fine art galleries in Nashville that provide an ideal forum for traditional and contemporary artists. Photo by Bob Schatz

Nashville At Home

Middle Tennessee communities are full of people creating and nourishing a sense of local pride and unity—a process that continues both in spite of and because of change. In that context, neighborhoods throughout the region are established, flourish, decline, and are reborn. Like the people within them, they have

lives and cycles of their own.

Mostly though, neighborhoods just work. People move in, get to know the territory, go to the corner grocery, play in the parks, rear their children, attend the schools and churches. They shape the character of the streets on which they live. They call it home.

LIVING WITH THE PAST

Nashville's earliest streets bordered the heavily trafficked Cumberland River. Merchants built two-story shops on busy Second Avenue (then called Market Street), sold their wares downstairs, and lived with their families in the upper stories. They formed a bustling neighborhood.

As the city grew, massive Italianate warehouses replaced those earliest riverside shops, but families still occupied the area and newer residential development spread in every direction. Today it is part of the city's urban core, but in 1850 the Rutledge Hill area (a few blocks south of Broadway on Second Avenue) was an industrious and elegant suburb. The same was true of Edgefield across the river and of

Germantown west of the capitol.

Early in this century, Germantown's Assumption Church was a strong Catholic congregation, but by 1979 it had become a small mission parish. The problem was the neighborhood. Over the years, Germantown had become increasingly commercial and industrial. Many longtime residents had moved away. Century-old houses were boarded up. The area was decidedly unsavory—reputed for drug deals, prostitution, and crime. Assumption's members felt isolated, even threatened. But change was on the way.

In 1979 a small but determined group of professionals bought several contiguous neighborhood houses and set about restoring them. They rewired and replumbed. They knocked out walls and replaced roofs. They signed a neighborhood compact, the beginning of a community association. They repaired brick sidewalks and cast-iron fences. They had the area designated as a historic district. They hung colorful banners from light poles and opened an upscale restaurant.

In 1981 the fledgling Germantown

The vital spirit of Nashville's many neighborhoods creates a special environment in which to live and work. Photo by Bob Schatz

Charming condominiums such as these on West End Avenue provide an alternative to the traditional single-family dwelling. Photo by Bob Schatz

Neighborhood Association started an annual Octoberfest celebration, a kind of homecoming for the community. The event has grown every year since. In 1988 the festival drew 16,000 people for an afternoon of music, home tours, sauerkraut, and dark beer.

After the turn of the century, expanded streetcar lines and the popularity of the automobile facilitated residential development farther from downtown. Grand estates on the edge of the city were divided into lots, streets were laid out, and houses appeared. The grounds of Belmont Mansion became a 1920s suburb and today's Belmont-Hillsboro Historic District.

Originally part of the grand Belmont estate, the oldest sections of the Belmont neighborhood were subdivided in 1890, and the newest sections during the 1940s. From the beginning, diversity has been a neighborhood hallmark.

The area contains a wide variety of architectural styles, and a large portion of the neighborhood is listed on the National Register of Historic Places. Bungalows, Four-squares, Tudor revivals, and International houses line Belmont's shady streets. Located near the university district (Vanderbilt University, Belmont College, and David Lipscomb College are all within walking distance) Belmont is also one of Nashville's more integrated communities. An active

Belmont: A Certain Patina

Photographer Bill LaFevor grew up in Nashville. Fifteen years ago he and his wife Chris bought a large 1920s home in the Belmont neighborhood. They and their three sons still live there today. LaFevor declares he wouldn't have it any other way.

"We love it here," he says. "But there are people who wouldn't want it. They just don't know the secret about living this close to downtown—that it's at least as quiet as living in the suburbs. Living here, you don't have to join that rush of traffic going into town each day. I've known lots of people who have moved out of town to get away from the noise and rush, but they just end up fighting the same thing further out.

"The interstate conflict was something else. We had a lot of noise during construction. Fortunately we hardly notice the road now. That's because

the neighborhood forced changes in design that made the thing livable. Ultimately, our protest made the interstate accommodate itself to the community instead of the other way around. In fact, my primary regret about the changes of recent years is not the interstate—it's the loss of the corner groceries and drugstores. We used to have one every few blocks, but they just can't compete with the big chains, so we've lost quite a few of them.

"These old houses were built during the days when labor was cheaper. People could afford those little touches like plastering and fine woodwork. These houses take on a certain patina as the years go by. I like that. You can't duplicate that effect in newer housing. The houses, the tall trees, the stone fences—they come together to make something very special."

Many of Nashville's young professionals have established their homes in the restored structures of historic Second Avenue, which have been converted into spacious lofts and roomy condominiums. Photo by Bob Schatz

neighborhood group has taken the lead in maintaining this ethnic and economic diversity.

Currently, Belmont faces the challenges of increased density and commercialization. Five years ago, Interstate 440 was built virtually through its center. Neighborhood groups fought the proposed interstate avidly, forcing substantial design concessions.

Once home to world-famous racehorses, the fields of Belle Meade Plantation were developed into a wealthy and prestigious community. Robert Weakley's East Nashville estate became the community of Lockeland Springs. You can't drink it anymore, but mineral water still runs at the old spring.

Newer but still historic neighborhoods are plentiful, including Richland-West End, Woodland-in-Waverly, Sunnyside, Woodmont East, Seventeenth District, Buena Vista, East End, and Old Hickory. The list goes on. In fact, if you highlight historic neighborhoods on a contemporary map of the city, they form a virtual ring around downtown. Nine Nashville neighborhoods are listed on the National Register of Historic Places. A dozen others offer urban amenities and homes

half a century old.

Characterized by ethnic and economic diversity, they are a varied lot. Some older neighborhoods are prime locations, their value enhanced by Nashville's recent growth and a strong regional preservation ethic. Others are poorer, the houses decayed and just beginning to see restoration. But in every case, the community is alive and active. Nearly 20 neighborhood organizations strive to improve the way of life in Nashville's oldest residential areas, and one thing is for certain—time is on their side.

Listed on the National Register of Historic Places, the Edgefield Historic District has undergone impressive renovation and restoration in recent years. Elegant Victorian homes, townhouses, and bungalows now reflect the sense of security and accomplishment felt by the district's residents. Photo by Bob Schatz

Germantown: Plenty Of Chiefs And Very Few Indians

Architect Michael Emrick's Victorian cottage was built in 1840. By 1979, when Emrick bought the house, the years had taken their toll. He spent two months making the structure habitable before he could move in. Even then, there was lots to be done.

"For a while," Emrick remembers, "I had to do my dishes in the bathtub. It was pretty grim. But even at its worst, I could see what the place could be. Now I have terrific living space in the back section of the house, and my offices are in the front two rooms. I've even opened the house for tours during Octoberfest. The whole arrangement works quite nicely—the house *and* the neighborhood—but both were a struggle.

"In a very real sense, we *took* this street. We reclaimed it from the drug pushers and the winos and the prostitutes who had made it unlivable for years. We made it a neighborhood again. A place where children can play. A place where everybody up and down the street knows everybody else. It's the kind of place where everybody should grow up.

"As soon as I moved in, I found terrific families in the neighborood. They had been here for decades. They welcomed us newcomers with

Photo by Bob Schatz

open arms. I got more or less adopted. If I got sick, chicken soup came over. That sort of thing.

"Restoring a whole community takes people who are either very secure or totally crazy. It takes people who are willing to work for something, willing to say 'we've found something we want, and we're going to make it work.' In a situation like this you end up with some very strong personalities—with plenty of chiefs and very few Indians. Personally, I wouldn't have it any other way."

IN THE 'BURBS

Postwar prosperity during the 1940s and 1950s effectively transformed the residential face of Nashville. Thousands of new residents moved to town during those years to take advantage of the area's growing business and industry, and a host of new suburbs developed to accommodate them. Outlying communities that had once been mostly fields, forests, and filling stations became thriving neighborhoods of subdivision, grocery stores, movie theaters, and shopping centers. They've been busy, prosperous places ever since.

The postwar expansion occurred all around town. West Meade, Green Hills, and Belle Meade developed to the west

of Nashville, Donelson and Hermitage developed near Andrew Jackson's famous country estate to the east, Madison expanded to the northeast while Crieve Hall and Antioch mushroomed to the south. The growth of areas outside Nashville's traditional limits transformed earlier, streetcar suburbs into city-center neighborhoods, and the newer, outlying areas became Nashville's classic "burbs" as we think of them today.

With the city's recent growth, density has increased somewhat in these postwar communities, but for the most part they remain quiet and established—neighborhoods characterized by winding streets, large green lawns, brick ranch houses, and colonial revivals. Forty years old now, these communities have developed a sense of continuity. A second generation of Nashvillians attend the same high schools from which their parents graduated, and trees in the yards have grown to maturity. Established when the areas were first developed, churches and retail businesses near the half-century mark. Right now Green Hills, Crieve Hall, Hermitage, and Donelson are essentially in the process of becoming old-timers, their newness displaced by Nashville's expansion further into the countryside.

REGIONAL FLAVOR
In 1976 writer Peter Jenkins walked across the country and wrote a book about his experiences. At the end of his adventure, he bought a farm in Tennessee. Jenkins chose a scenic little community pretty much off the beaten path to anywhere, about 40 miles south of downtown Nashville—a place called Spring Hill. Then in 1985, General Motors announced its choice of location for an enormous new auto plant that would produce its model Saturn. They, too, had chosen a scenic little community pretty much off the beaten path to

anywhere . . . Well, you get the idea.

Jenkins was not alone in his surprise. Plenty of Middle Tennesseans who used to enjoy the quiet of rural life suddenly found themselves in the midst of growth and change.

By the time the city of Nashville and Davidson County became a single metropolitan government in 1963, residential development had penetrated almost every corner of the county. Then during the 1970s, five of the six counties surrounding the city began to flourish as bedroom communities, brimming with residential developments that catered to commuters. And during the 1980s the effects of growth were felt even further out, in counties once removed from Davidson and beyond. Today half a million people live in Nashville proper and half a million more make their homes in the outlying community and countryside.

North of downtown in Sumner County, Hendersonville and Gallatin (once separate towns with agrarian economies) have become home to many stars of the music recording industry. They've also become enormous bedroom communities. At least 14,000 Sumner Countians commute into Davidson County to work each week-

The annual Germantown Octoberfest, founded in 1981 by the Germantown Neighborhood Association, has become a popular Nashville event, instilling the community with a sense of spirit and pride. Photo by Bob Schatz

Residential construction exploded in the outlying regions of Nashville during the postwar boom of the 1940s and 1950s, establishing suburban communities for the increased population. The Green Hills area has developed into a quiet neighborhood with a strong sense of community. Photo by Bob Schatz

ABOVE: A rural atmosphere is still prevalent in the region, as illustrated by the existence of businesses such as Earl's Fruit Stand in Franklin. Here, Earl strikes a proud pose with his pumpkin harvest during the colorful autumn season. Photo by Matt Bradley

RIGHT: Tree-lined avenues and an array of architectural styles characterize the neighborhood of Belmont, situated just 10 minutes from downtown Nashville. Photo by Bob Schatz

day. To the east, in the Wilson County town of Mount Juliet, 100-year-old farms and brand new houses sit side by side in the evolving landscape. To the south, in Rutherford County, Smyrna is home to Nissan U.S.A. New subdivisions and shopping areas serve the auto plant's 3,200 employees, and a host of other, smaller businesses and their employees have moved to the area to support and supply both the Nissan and Saturn operations.

In many cases, life in these formerly rural areas still has a small-town atmosphere. Farmers still drive their pickups and flatbeds into town to shop. But folks remember when those same farmers used to bring in produce and chickens once a week to sell in the town square. They remember when the courthouse square was where the community's tobacco-chewing retirees used to sit and whittle away their afternoons. But Middle Tennessee's county

Franklin: Small-Town Flavor, Big-City Issues

Jan Tabernick is a documentation specialist, and her husband, Ty, a software consultant. Three years ago when Ty's company transferred them and their two daughters from Michigan to Nashville, they had just one week to find a house. In the midst of their frantic search, Jan and Ty checked out a new subdivision just south of Franklin. They drove up a wooded hill to a home under construction, climbed the steps onto the front porch, and took in the view. Green, sunny fields bordered by stone walls ran to the horizon. Their search was over.

"Housing prices were higher here than I expected," recalls Jan. "Of course, Michigan was depressed at that time and Tennessee was booming. So this house was more expensive than I thought we could afford. But after we saw that view, we were sold. After that, nothing else compared.

"We wanted to feel like we were 'out.' We wanted woods and space. But I work near downtown Nashville. I couldn't face a terribly long drive every day. This is a terrific compromise.

Photo by Bob Schatz

It feels like the country, but we're only five minutes from the interstate.

"The whole Franklin area is like that. You've got big-city living close by. I mean we're 45 minutes, maybe less, from the airport. But at the same time Franklin still retains a certain quaint atmosphere. It's really enjoyable. I love going into 'town'—Franklin I mean."

---

seats are yuppier these days, and the pace isn't nearly so leisurely anymore.

Residential real estate in little towns like Franklin, Clarksville, Dickson, and Centerville has increased in value as these communities have swollen with their shares of Nashville-bound commuters. All totalled some 60,000 people commute to work in Nashville each day.

Between 1970 and 1984, the population of Williamson County doubled. It became the fastest-growing county in the state—and still is. In Franklin, the county seat, a central square lined with turn-of-the-century buildings and neighborhoods of Victorian houses belie the modern town currently taking shape. Investors have transformed downtown Franklin into an elegant mix of restaurants, shops, offices, and apartments.

And in every direction outside town, what was recently rural countryside is now 1990s suburbia.

Not only is Franklin getting bigger, it's getting richer. The median value of a home sold in Williamson County these days is more than $70,500—the highest in the state. Twenty years ago the value was just $17,000. Franklin's per capita income is 10 times greater than it was 20 years ago, jumping from $3,400 in 1970 to $35,000 today. In Franklin, the times are a-changin' for sure.

Such rapid growth has produced hodgepodge development. Fast-food operations and gas stations at the interstate exit into Franklin offer no hint of the historic district two miles away. Recent local elections have been fought primarily around growth-management

issues. In the end, Franklin's biggest challenge during coming years is bound to be preserving the very qualities—the quiet and charm—that attracted so many people there in the first place.

For people in towns as far away as Columbia to the south and Portland to the north, Nashville is the major metropolitan center—the place to go for concerts or shows, for specialty shopping, or to catch international flights. Interstates entering the city from every direction put more than one million residents within less than an hour's drive of downtown.

Bedford County's seat is Shelbyville, traditional home of the Tennessee Walking Horse. The annual Walking Horse Festival draws upwards of 235,000 people to Shelbyville each year. Pulling highly bred walkers in trailers behind their trucks, horse owners crowd the town square, fill the motels, and wait in line at Pope's Cafe, where the apple pie is homemade and a meat-and-three plate still costs less than four dollars.

During the rest of the year, Shelbyville and Bedford County support 29,000 people, a significant number of whom commute the hour or so to Nashville to work Monday through Friday. Those folks mostly stay in Bedford County on weekends, enjoying the peace and quiet of the countryside, working their gardens, and attending country churches. But they may come to Nashville for dinner and a show. Many shop at Hickory Hollow, the Nashville mall nearest them.

Unlike those counties immediately adjacent to Davidson, Bedford has not experienced much of Nashville's spillover growth—yet. But most people realize that it's just a matter of time.

Clearly, Nashville is no longer the big town it was during the 1940s. Nor is it the small city it was during the 1960s. For all practical purposes, Nashville has expanded to take in the whole Middle Tennessee region—a region encompassing log cabins and skyscrapers, horse farms and courthouse squares, condominiums, Victorian cottages, and neighborhoods of every description.

Bedford County:
Fourth Of July Fireworks And Trains Passing Through

Noel Johnson manages a hardware store, is a volunteer fireman, and is a member of the school board. His wife, Wanda, is an artist and historic preservation consultant. They live with their two children in Wartrace, seven miles outside Shelbyville. In the summer, they bicycle the roads outside Wartrace and canoe the Duck River, floating past the limestone foundations of abandoned nineteenth-century gristmills. Their Italianate vernacular house has 18-foot ceilings and fireplaces in every room. It is not, Noel admits, the kind of house they could afford in downtown Nashville.

"I grew up in Wartrace," he recalls. "We've talked from time to time about moving, but we choose to stay. We're here because we want to be. There's lots we like about the place. For one thing, we live here—and live well—for less than we could in Nashville proper. We're 50 miles down the road from downtown Nashville, but with interstates the way they are, we're as close to the airport time-wise as lots of Nashvillians are. Besides that, we've got the peace and quiet. About the only noise in Wartrace is Fourth of July fireworks and trains passing through."

Quality Learning, Quality Caring

Nowhere is Nashville's innovative spirit more evident than in its commitment to the education and health of its residents. Throughout the city's schools and hospitals there are shining examples of the care and nurturing that sets this city apart from others. Many of the special programs and facilities used here have

served as prototypes for those in other communities.

PREPARING FOR TOMORROW
When Nashville's founding families arrived on the banks of the Cumberland River at the site of the city's first settlement, they lost little time in establishing a school for their children. That tradition of commitment to education prevails in Nashville today.

Nashville is called the Athens of the South because it is home to more colleges and universities, 16 in all, than any other city in the United States, except for New York City. Vanderbilt University, with its prestigious law and medical schools, is located here. So is Fisk, the first predominantly black institution to be awarded university status. Meharry Medical College, where more than 40 percent of our country's practicing black physicians and dentists have earned their degrees, is another well-known Nashville school.

PUBLIC EXCELLENCE
The city's 120 public schools are consistently recognized at state and national levels for their successful and innovative educational programs. The city boasts eight schools recognized as national "Schools of Excellence" by the U.S. Department of Education. Nashville's students regularly score above state and national averages on achievement tests. The faculty is well-qualified—62 percent of the teachers have at least a master's degree.

Former Tennessee governor Lamar Alexander's "Better Schools Program" laid the foundation for many of the good things that are happening in Nashville schools today. Alexander set new standards for academics, computer literacy, and teacher certification in the state. The program also provides economic incentives to keep the best teachers on the job.

Nashville classrooms stress basic skills first. At the elementary level, reading

A wide selection of medical services combined with a strong educational system provides the Nashville area with a healthy and well-educated style of living. Pictured here in the comfortable surroundings of the birthing unit at Southern Hills Medical Center are two proud parents and their new child. Photo by Bob Schatz

61

With the active assistance and unflagging faith of teachers and parents, Percy Priest Elementary School now enrolls more than 480 children in its kindergarten through fourth grades. The school's new computer lab, paid for through a fundraising drive, helps to introduce the students to the new world of high technology. Photo by Bob Schatz

and math comprehension are emphasized. Seventh and eighth graders are required to complete a computer literacy course. High school students must pass a proficiency test before receiving their diplomas.

The gifted and talented in Metro schools also benefit from programs designed especially for them. Three schools for high academic achievers have been opened. One of these, Meigs, serves more than 600 middle-school students, while the Martin Luther King Jr. School for Health, Sciences and Engineering has 400 seventh through twelfth graders enrolled in its special programs.

Hume Fogg Academic, the alma mater of songstress Dinah Shore, now teaches 480 of the city's brightest high school students. The students selected to attend these schools pursue independent studies in addition to their highly concentrated curriculums.

Enrichment programs are available throughout the Metro school system. All students in grades kindergarten through eight participate in the EXCEL program, which provides exposure to a wide variety of subjects not included in the day-to-day curriculum. The ENCORE program, designed for the gifted and talented, offers advanced study for qualified students through the eighth grade. Gifted students at the high school level may enroll in the Scholar's Program, a demanding curriculum in hard-core academics and advanced placement studies. The Extended Learning Program offers noncredit, after-school classes in subjects ranging from sculpture to music video production. These courses are open to artistically and academically talented junior and senior high students.

Six public schools provide for the special needs of children with physical or mental handicaps, but many special education students are mainstreamed. The Bilingual/English as a Second Language Program serves about 700 students from 53 countries representing 44 native languages. A state-funded pilot program to reduce the dropout rate is currently under way. The Learning Center, located in the Preston Taylor housing project, provides tutoring, parenting skills training, and adult literacy classes. It is the first in the nation and has been highly praised by the U.S. Secretary of Education. A second nationally recognized program, the Caldwell Early Childhood Education Center, helps disadvantaged parents prepare themselves and their preschool children for elementary school.

In addition to academic programs, students at the 10 comprehensive high schools can receive job training in more than 30 fields ranging from aircraft

mechanics to photography, from computers to plumbing.

While about 15,000 Nashville children attend private schools, the vast majority, more than 67,000, attend public schools. Since 1984 public-school enrollment has steadily increased. Joyce Vise, the community relations representative for Metro schools, attributes the increase not only to the quality of education the public schools provide, but also to community support and involvement. Last year more than 4,000 interested parents and citizens volunteered over 176,000 hours to the schools

through Metro's Good Friends Volunteer Program.

"I say a prayer of thanks every night for my school parents," says Principal Dorothy Butler as she walks through the halls of Percy Priest Elementary School. "They're a committed bunch and I couldn't run this place without them."

Although it is several days before school opens, a number of parents are at the school working on projects. Some are decorating the halls, others are putting together the school handbook; several are helping teachers prepare their classrooms. "I don't even have to call them," says Butler as she greets each parent. "They just come in and do whatever needs to be done."

Percy Priest parents have a long tradition of school involvement. Opened in 1957 and named for a former congressman, the school quickly established itself as a community center. Lucy Vorhees, who served as an early PTA president, remembers that "everyone in the neighborhood sent their children there. Any neighbor you needed to see you'd probably run into at the school. It was that kind of place."

EXCEL is an educational enrichment program that offers exposure to subjects not included in the day-to-day school curriculum. Photo by Bob Schatz

These young scholars are just a few of the more than 67,000 children attending public school in the Nashville area. Photo by Bob Schatz

Parents were encouraged to take active roles. Whenever a teacher needed a helping hand, a line of parents were waiting, anxious to help. "The school was so small it didn't even have a telephone in the office," says Vorhees. "The parents got right together and organized a fundraiser. I think it was a bake sale. We made around $100, which was enough to pay to have the phone installed."

Throughout the 1950s and 1960s enrollment continued to climb. But when the 1970s brought integration and busing, things began to change. Many residents of the upper-middle-class neighborhood, where the school is located, began sending their children to private schools. Enrollment at Percy Priest steadily declined. One year there wasn't even a kindergarten at the school, and the first and second grades were combined into a single class.

Still, "the faculty and parents who remained didn't give up," remembers Becky Long who, along with her husband, Bill, played an active role in the school's revival. "The parents, the staff, everyone beat the bushes for students." The parents prepared a brochure and sent it out to everyone in the district. They called friends and neighbors and told them about the school. They even hosted open houses for the community at the school.

When Dot Butler joined the staff as principal in 1982 there were just over 200 students in the entire school. "There was unflagging school spirit though," says Butler. "I felt it when I first arrived. I still feel it." A veteran educator, Butler's energy, savvy, and marketing skills meshed perfectly with the school's contingent of dedicated parents. "Dot's arrival provided our school with an effective champion," says Long. "She took our case to the school board and won some battles for us."

In 1983 a citywide school redistricting plan helped stop Percy Priest's enrollment slide. Another nearby elementary school closed and Percy Priest inherited many of its students. Since then enrollment has steadily increased and the school has reclaimed many of the neighborhood children who had switched to private schools. Today more than 480 children, in kindergarten through fourth grades, are enrolled at the school.

As the school has grown, parental involvement has increased. The PTA Board now oversees a sophisticated committee operation that covers 20 areas of school life including publicity, volunteers, government affairs, and long-range planning. The PTA pays for the school music teacher as well as the art teacher. Money for a new computer lab was obtained through a school-wide fundraising drive. In this racially balanced, middle-class school district, the PTA has little trouble raising more than $40,000 annually.

"Our goal is to get every parent involved in something at the school," explains Missy Russ, who has served as PTA president and fund-raising chairman for the school. "We believe there's a correlation between the level of parental involvement and the quality of the school."

Businesses are actively involved with schools through the PENCIL Foundation and its Adopt-A-School program. When Eakin Elementary experienced traffic and parking problems at its facility, the school's Adopt-A-School partner, Gresham and Smith and Partners, sent over a traffic engineer who developed plans to solve the congestion.

When students at Harris-Hillman School needed rides to a special Sesame Street production, the school principal simply put in a call to its Adopt-A-School partner, WKRN-Channel 2, which arranged for buses and provided tickets to a special afternoon performance.

The Pencil Foundation's Adopt-A-School program enables private businesses to get involved with the city's 120 public schools. Founded eight years ago, the program was quickly

embraced by the business community. Today more than 170 businesses are involved.

"Adopt-A-School has helped change the business community's perception of public schools," says Cavert School principal Rob Sasser. "Through this program business leaders get to know principals, teachers, and students first-hand. The businesses don't just send awards or money over to the schools. The one-on-one experience is very im-portant to them."

There are few formal guidelines for the schools and their partners to fol-low. "Because so many different sized businesses and schools are involved, every partnership is different," explains Pat Wallace, program director. "Some businesses provide money and supplies to the schools, others offer manpower. The relationship is limited only by imagination."

Rob Sasser enjoys that flexibility.

The Diet Guru

In 1986 more than 30,000 Nashvillians went on the Rotation Diet and shed over one million collective pounds. This Melt-a-Million campaign was spearheaded by Dr. Martin D. Katahn, a weight-management expert at Vanderbilt University.

Grocery stores reported runs on diet-approved foods. Restaurants began offering Rotation Diet specials. Thousands of people participated in weekly weigh-ins sponsored by a local super-market chain.

News of the Nashville success spread quickly across the nation. Katahn's book, *The Rotation Diet*, catapulted onto best-seller lists everywhere. Across the country, "Are you rotating?" seemed to replace all other conversation.

A Julliard student turned psychology profes-sor, the soft-spoken Katahn seems the unlikely choice to whip millions of Americans into shape. But to his legions of followers, Martin Katahn is America's diet guru—Richard Simmons and Jane Fonda all rolled into one.

Katahn brings a decidedly personal view to weight loss. He was a fat kid in a fat family. "When I was little I couldn't bend over in my snowsuit to make snowballs. By the time I was 12, I was 50 pounds overweight."

In 1963, at age 34, he suffered a heart attack while playing ping-pong. "At that point I either had to change my eating habits and lose weight, or die." Although he didn't know anything about nutrition back then, Katahn developed a pre-cursor of the popular Rotation Diet for himself.

Photo by Bob Schatz

"I'd diet and exercise like hell for three weeks, then take a short break from the diet while I con-tinued to exercise." It took him 18 months to drop from 230 to 154 pounds, where he re-mains today.

The wiry, 60-year-old Katahn says he main-tains his weight by practicing what he preaches. "I don't diet; I eat sensibly." For exercise he walks, jogs, and plays a mean game of tennis.

"In our partnerships with Touche Ross and the Independent Order of the Odd Fellows, I'm not shy about asking them to help out whenever and wherever we need it." Cavert's partners have generously provided special playground equipment for the school, given Christmas gifts to all the children, and hosted a Teacher Appreciation luncheon.

Sasser says one of his most meaningful partnership experiences occurred when he was at the Murrell School.

Because Murrell is a special education school, the typical partnership things like incentive awards weren't appropriate. Instead the school's partner, IBM, funded an artist who came to the school and helped the students work in clay. "We had an art show and invited all the parents. It was the first time many of the parents had seen any arts and crafts work done by their child. It was very moving to see the parents' reaction."

In addition, two local private foundations, Hospital Corporation of America (HCA) and the Metro Nashville Public Education Foundation, provide grants to individual teachers to pursue creative teaching and learning experiences.

More than 57 percent of Nashville's high school graduates pursue advanced studies or degrees. Because the city is blessed with an abundance of high-quality colleges and universities, Nashville's "best and brightest" can earn advanced degrees right in their own hometown.

VANDERBILT UNIVERSITY

Perhaps the best-known school in town is Vanderbilt University. Founded with a million-dollar gift from Commodore Cornelius Vanderbilt, the university has been influencing American culture for more than 100 years. Its graduates include Pulitzer Prize-winning author Robert Penn Warren and heart transplant specialist Norman Shumway. The private institution offers undergraduate degrees in more than 50 fields and is widely known for its advanced academic offerings in medicine, law, business, and education.

Many members of Vanderbilt's faculty and staff are deeply involved in research projects. The university receives government and corporate grants to conduct studies in everything from biochemistry to space exploration. Recently the university was selected as a space research center for NASA,

LEFT: Founded with a generous gift from Cornelius Vanderbilt more than 100 years ago, today's picturesque Vanderbilt University is conveniently located near downtown Nashville. Photo by Matt Bradley

BELOW: Tennessee State University students concentrate on their studies at the North Nashville campus library. Photo by Bob Schatz

Shopping For Health

James Sheer, nutritionist and dietician, is standing in the middle of a grocery store, a soft drink in one hand and cranberry juice cocktail in the other. He turns and asks, "Which contains more sugar, the soda or the juice?" Remembering my mother's exhortation that drinking soda will "rot your teeth," I point confidently to the soda. Sheer smiles his "gotcha" smile and then explains that cranberry juice cocktail actually has more sugar in it than most soft drinks.

Along with nine other health-minded souls, I'm attending a Savvy Shopper Supermarket Tour led by Sheer. The tour, offered several times each month through Baptist Hospital's Center for Health Promotion, is one of many community outreach programs and services sponsored by the hospital.

For the next two hours, Sheer leads us up and down the grocery aisles, points out nutritious foods, and teaches us how to read food labels. Along the way he dishes out a generous helping of food facts like "Dark green vegetables are more nutritious than light green ones," and such helpful homilies as "Fresh is best."

Teaching a nutrition course in a grocery store really makes a lot of sense. We learn how buying the right foods can help cut fat, reduce cholesterol, control sodium, and add fiber. "A lot of people have trouble applying what they learn in regular nutrition courses to the real world, the world of the food store," explains Sheer.

Most of the people who take the shopper tour are interested in weight management—although the tour includes important information for diabetics and heart patients as well. "Who would like to lose 10 pounds?" Sheer casually asks. We all raise our hands. He leads us over to the spice section and grabs a bottle of a popular butter substitute. "Use this on popcorn, potatoes, almost anything that calls for butter, and I promise you'll lose 10 pounds in a year." We each put a bottle in our grocery carts.

Photo by Bob Schatz

"There's a healthy substitute for almost everything you like to eat but really shouldn't," says Sheer. "Pretzels can replace potato chips. Yogurt can be substituted for sour cream, margarine for butter, and sherbert for ice cream."

At the end of the tour our grocery carts are brimming with healthy, low fat, low sodium, low anything that's not good for us foods. On the way to the checkout lane I stop by the ice cream freezer. Throughout the tour I've willingly tossed fresh fruits and vegetables into my cart. I've passed up steak, avoided the candy aisle, and resisted all the great convenience foods. But can I pass up my favorite all-natural gourmet ice cream?

As the others in my shopper class watch, I patiently read the label—milk, cream, butter, sugar—a dietary delight. Based on what I've just learned I know that there's very little in this ice cream to recommend it nutritionally. But I buy it anyway.

one of only five such centers in the country. University researchers will receive millions of dollars for work that ranges from testing robots for the planned space station to making near-perfect metal in zero gravity. The Vanderbilt faculty has two Nobel Prizes to its credit, both in medicine. In 1971 the late Dr. Earl Sutherland, Jr., received the prize for his research in hormones. In 1986 Dr. Stanley Cohen received a Nobel Prize for his work in cell growth.

FISK UNIVERSITY

Fisk University is the oldest existing institution of higher learning in the city, founded just after the Civil War by Union General Clinton Fisk. Fisk is a private, four-year school designed to meet the special needs of minority students. In 1930 Fisk was the first black university to be accredited by the Southern Association of Colleges and Schools. In 1933 it was the first black school to be placed on the approved list of the Association of American Universities.

Fisk counts among its graduates W.E.B. DuBois, cofounder of the NAACP; George E. Haynes, cofounder

of the National Urban League; and poet Nikki Giovanni. The 42-acre Fisk campus is a National Historic District.

Fisk University is the home of the Jubilee Singers, a nationally recognized singing troupe composed of students at the school. The group was formed by ex-slaves soon after the school opened. The singers began touring in 1871; they traveled north and earned $25,000 for the school. They soon gained international acclaim, making two tours of Europe and raising another $100,000. The money was used to purchase the present campus and to construct Jubilee Hall, the first structure in

ABOVE: A private four-year insitution, Fisk University was established in the years following the Civil War, and in 1930, was the first black university to be accredited by the Southern Association of Colleges and Schools. Photo by Bob Schatz

ABOVE LEFT: More than 100 major fields of study are offered to the students of David Lipscomb College. Photo by Bob Schatz

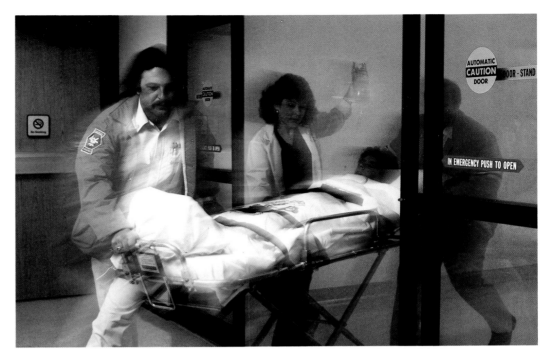

this country built specifically for the education of blacks.

TENNESSEE STATE UNIVERSITY

Tennessee State University (TSU), a branch of the state-operated university system, maintains two campuses in Nashville for its 7,000 students. Its downtown facility caters to working students by offering many late-afternoon and evening courses, while the school's main campus is oriented to full-time students.

ADDIITONAL HIGHER EDUCATION

Nashville is home to several church-related colleges and universities representing various denominations. David Lipscomb College, a four-year school supported by Church of Christ denominations throughout the country, traces its beginnings back to 1891.

More than 2,000 students attend classes at its 65-acre campus on Granny White Pike. Seventeen departments offer more than 100 major fields of study.

Belmont College, a private four-year Baptist school located near Music Row, was one of the first colleges in the country to offer a degree in Music Business Administration. Students from all over the country who want a career in the record industry have enrolled in its specialized courses, ranging from record promotion to studio engineering.

ADVANCED DEGREES

Two distinguished medical schools are located in Nashville.

The Vanderbilt School of Medicine is part of Vanderbilt's massive Medical Center complex. The school boasts more than a dozen research centers including cancer, heart disease, clinical research, and molecular pharmacology. The Diabetes Research Center, the first established by the National Institutes of Health, includes physician training and research in the diagnosis and treatment of diabetes.

About 40 percent of the black physicians and dentists practicing in the United States today were educated at Meharry Medical College. Almost 75 percent of the graduates practice in medically underserved inner cities and rural areas. Comedian Bill Cosby recently contributed nearly one million dollars to the school to fund a teenage pregnancy

prevention project. The school also supports research in sickle cell anemia, kidney disease, and hypertension—all of which strike Meharry's target population with particular severity.

Of course, many medical school graduates stay right here. Over 2,240 doctors practice in Nashville, more than in any other city in Tennessee. They join a highly competitive market where health care dollars are aggressively courted.

MEETING TODAY'S NEEDS
Competition forces hospitals to constantly experiment with new programs and technologies. "No hospital can rest on its laurels in this town," says Vickie Beaver, director of the Women's Health Services Department at West Side Hospital. "It's an open market here."

Hospitals race to offer the latest technologies and treatments to patients and doctors. Throughout the city billboards tout special hospital services. Radio and television ads abound, featuring popular personalities like Barbara Mandrell and Bill Cosby promoting their favorite Nashville hospitals. Mass mailings entreat residents to attend a myriad of prevention and wellness classes offered through the hospitals.

This competition has pushed the city into the forefront of the national health-care industry. Over the last decade the city has emerged as a center for medicine and medical research, nationally recognized for its hospitals, physicians, and medical schools as well as for an unusually high number of specialized units that deal with specific health-care problems.

The Nashville area is served by more than 20 hospitals. Vanderbilt, St. Thomas, and Baptist are the three largest. West Side and Southern Hills are representative medium-sized hospitals, which have successfully made a place for themselves in Nashville by providing specialized health-care services like sports medicine and cancer treatment. More than a dozen 200-bed community hospitals also serve the area.

LEADING THE WAY
Vanderbilt University Hospital is probably Nashville's best-known health-care facility. Recently featured in the book, *The Best Hospitals in America*, the 661-bed university hospital boasts some of the most innovative research, training, and specialized health-care facilities in the country.

Vanderbilt University Medical Center offers state-of-the-art equipment and technology at its 661-bed facility. Photo by Matt Bradley

The hospital's Center for Transplantation, which opened in June 1989, is the only one of its kind in the United States. The center coordinates the activities of the hospital's transplantation teams—heart, heart/lung, pancreas, kidney, and bone marrow. "Many universities do different transplants on their campuses, but almost all are done in self-contained groups," says center co-director Dr. J. Harold Helderman. "At Vanderbilt we're providing the environment so that all our transplantation people can work together."

Under the center's auspices Vanderbilt doctors train St. Thomas medical personnel in kidney and pancreas transplantation. "We're committed to putting together a group of interested physicians, scientists, and ancillary people whose goal is transplantation with a capital T," says Helderman. "As partners working together we can foster our goals."

Vanderbilt Hospital boasts one of the few Cooperative Care Centers in the United States. Here family members are trained by the medical staff to perform many of the services traditionally handled by nurses during hospital stays. Its Parkinson's Disease Center is one of

only six such centers nationwide.

The Vanderbilt Children's Hospital is the prototype for other children's hospitals around the country. Focusing on the specific problems of childhood and adolescence, the doctors and nurses here care for babies born with birth defects, children with diabetes, and hundreds of other young people with health problems.

Baptist Hospital, a full service non-profit 730-bed facility located just two miles from the Vanderbilt complex, is the largest healthcare facility in Middle Tennessee. The hospital offers a broad range of services including physical rehabiltation, a specialized Heart Care Center, a cardiovascular diagnostic lab, and a trauma treatment center. Its Women's Pavilion is devoted entirely to women's health care including ob/gyn, breast surgery, and menopause treatment. The center offers a host of medical workshops including prenatal care, child safety, and birth control. Baptist introduced overnight surgery service to Nashville over ten years ago. Today the hospital boasts a separate facility for same-day surgery.

St. Thomas Hospital, one of the oldest hospitals in the city, is a 571-bed,

The technologically advanced lithotripter at Baptist Hospital provides an alternative method for treating kidney stones, sometimes alleviating the need for traditional surgery. Photo by Bob Schatz

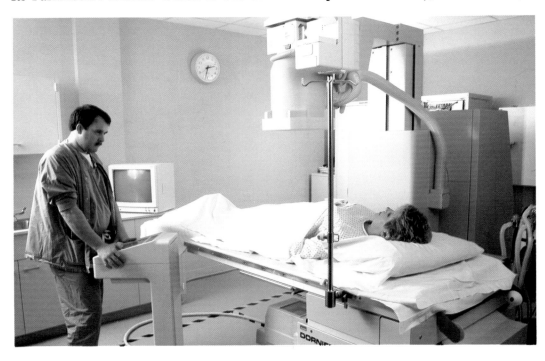

acute-care, teaching and referral hospital operated by the Daughters of Charity. Established in 1898 by the Catholic Religious Order of Women, St. Thomas is a leader in cardiology and cardiovascular surgery. The first coronary balloon angioplasty in the region was performed here in 1979, and the first heart transplant operation in Tennessee was performed here in 1985.

Seventeen other hospitals throughout the area provide care for Nashville's growing population. Outlying communities are well served by community hospitals like the University Medical Center in Lebanon and the Sumner Memorial Hospital in Gallatin. Closer to town, West Side and Southern Hills have each emerged as strong niche marketers in specialized fields—ob/gyn for West Side and sports medicine for Southern Hills.

As Nashville enters the 1990s, the variety of education and health care programs and facilities available here will help prepare its residents for the new century. From educating its young to caring for its old, Nashville sets a standard envied by many.

All About Birthing Babies

For years Nashville was a traditional town when it came to delivering babies. There were a limited number of private rooms available on maternity wards. Fathers were permitted to be involved in deliveries only if they attended special classes. Sibling visits were limited to special hours. But in 1985 West Side Hospital opened its Family Birth Unit and changed the Nashville birthing business forever.

A small, 200-bed facility, West Side was searching for a niche in the highly competitive Nashville medical-care market. Maternity seemed a natural choice, since the hospital had already established a name for itself in women's health. However, the maternity field was dominated by two large hospitals—Vanderbilt and Baptist. To compete effectively, West Side needed to offer a unique approach to the birth experience.

"We wanted to offer a maternity experience that complimented our other services," explains Vicki Beaver, who served as a consultant on the project and is now director of women's health services development for the hospital. "Since West Side is a small, family-oriented facility it seemed natural for us to establish a family-centered maternity ward."

What West Side ultimately developed was a maternity unit composed entirely of private, hotel-like labor-delivery-recovery-postpartum (LDRP) rooms, which enable the mother to experience

Photo by Bob Schatz

birth in the privacy of her own room.

West Side's birth unit resembles a comfortable, upscale hotel. Each room is large, private, and carpeted. Mauve and taupe replace the hospital-issue green. Rocking chairs, VCRs, and jacuzzis are standard—as is a champagne dinner.

The entire family-birth unit is so unhospital-like that it's hard to imagine that births actually take place here. But they do. Last year 2,000 babies were born at West Side, making the hospital the second-busiest birth center in the city. The unit's success affects the entire hospital's operation. "Maternity is a gateway service," explains Vicki Beaver. "If a family has a good experience in maternity, they'll come back for the other services we offer."

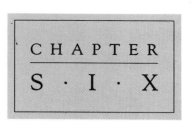
Year-Round Fun

One June morning in 1881, all trading on the New York Stock Exchange stopped for a moment and spontaneous celebration broke out. The cause of the jubilation came from Epsom Downs in England. Iroquois, a three-year-old thoroughbred from America, had won the English Derby. He was the first United States horse ever to

win the derby, a victory that would not be repeated for a long 73 years.

After his winning run, Iroquois retired to Nashville to spend the rest of his pampered life standing stud at the Belle Meade plantation, one of the most renowned thoroughbred farms of its era. Nashville has never forgotton that horse.

Each year in May on the Saturday after the Kentucky Derby, more than 40,000 spectators gather in a Nashville park just a few miles from the champion's paddocks. There they watch some of the world's finest horses run for the richest steeplechase purse in America. It is the nation's oldest continuously run amateur steeplechase, and they call it the Iroquois.

Over the past 45 years, the Iroquois has become much more than just a day at the races. The event is Nashville's unofficial rite of spring, attracting beer-toting college students, who watch from the grassy hillsides, as well as attracting the tonier set, who sip champagne in

box seats. Included in the Iroquois are eight races, among them a flat race, pony races, and two steeplechase races. Along with these sporting events, a social weekend has evolved that draws people from all over the country.

According to local legend, the idea for a race came up one morning in 1936, when Nashvillians Marcellus Frost and John Sloan, prominent horse owners, were enjoying a ride on one of the bridle paths in Nashville's Warner Parks. As they reined in their horses at the crest of a hill, Frost surveyed a vast meadow that sloped gracefully below them.

"You know," he remarked to Sloan, "if we could round up the money to put in two culverts, we'd have a perfect race course here." They soon pushed the project through. The course, eventually built as a three-year Works Progress Administration project, saw its first race in 1941.

At that point, the Iroquois event was new to Middle Tennessee, but horse

From the pounding of horses hooves thundering over the countryside in the Iroquois Steeple Chase to a refreshing walk through a landscaped park, Nashville is alive with year-round fun. Photo by Rudy Sanders

ABOVE: *The Iroquois Memorial Steeplechase is a time-honored springtime tradition. Photo by Bob Schatz*

ABOVE RIGHT: *Children of all ages delight in the many equestrian activities available in the Nashville area. Photo by Bob Schatz*

racing was not. Like central Kentucky to the north, Middle Tennessee has been prime horse country for nearly 200 years. By the mid-1800s, Nashville was recognized as a premier racing city with four tracks, one on each side of town. The richest horse race in the world to that time took place in 1843 at the Nashville Race Track. The race was called the Peyton Stakes, and the winner, a chestnut filly named Peytona, went on to win a celebrated match race in New York two years later in a contest billed "the race of the century." During the century since, both horses and riders from the Nashville area have made international news.

The big news during recent years, however, has been the Royal Chase, for while horse racing and steeplechase are hardly novelties in Nashville, royalty is. As a result, the turnout was impressive,

some 7,500 people attended in 1987 when England's Princess Anne came to ride the ponies in Nashville. The event turned into a sort of autumnal Ascot as well-heeled and hatted royalty watchers and horse lovers alike paid between $500 and $10,000 to occupy the box seats below the princess' pink and purple striped pavilion. Country music star Lee Greenwood sang "God Bless America," and Princess Anne finished third in her race.

While steeplechase has the highest profile among various horse events, there are plenty more in the Nashville region. All over Middle Tennessee people don English helmets and take riding lessons or canter along quiet bridle paths. They train Tennessee walking horses, hunt fox from atop tall jumpers, and even play polo on stout ponies especially suited to the sport.

It is no wonder, really, that Nashvillians enjoy their horses. Middle Tennessee is a landscape perfectly designed for equestrian pursuits and lots of other sport besides. The climate is warm three seasons of the year, encouraging outdoor fun. Rolling hills provide a perfect backdrop for year-round hiking, camping, biking, ball-playing, and jogging. Even more importantly, nestled amid those hills are waterways big and small, which are Nashville's primary playgrounds.

GETTING WET
You can't talk about warm-weather recreation in Middle Tennessee without talking about water. Davidson County is bordered by two lakes, Old Hickory to the north and Percy Priest to the southeast. The Cumberland River runs through downtown, and the Harpeth and the Buffalo rivers are within 45 minutes of downtown Nashville. Encircled by popular walking trails, spring-fed Radnor Lake is within the city

limits, and cold, clear Center Hill Lake is only an hour's drive away.

Old Hickory and Percy Priest offer miles of peaceful, accessible shoreline, perfect for the serenity of sailing or the fast pace of powerboats and water skis. A series of public docks houses nearly every kind of freshwater craft—canoes, flat-bottom fishing boats, luxurious private yachts, sailboats, and houseboats that can be rented for afternoon parties or long, lazy weekends. You can see them all at the city's annual Fourth of July party when the Cumberland is packed with a virtual armada of watercraft. While you're boat-watching, look for the opulent *Dolce Vita*, a 107-foot yacht belonging to local restaurateur Mario Ferrari.

Still Fishin' After All These Years

Talk show host Jimmy Holt sips his coffee, hunkers over in good-ole-boy fashion and greets former president Jimmy Carter. "We're just tickled to death to have you here," grins Holt.

The ex-president is on tour promoting his new book about the outdoors, and he's probably "tickled" too. After all, he has chosen the perfect vehicle for reaching area hunters and fishermen. On public television station WDCN for 18 years now, Jimmy Holt's *The Tennessee Outdoorsman* is the channel's most popular program. But not because it sends the blood racing.

Every episode is pretty much like every other. In his drawling, old-Nashville accent, Jimmy and co-host welcome a sportsman, usually someone from the region, to sip coffee and sit on the sofa with Jimmy. The three of them chat about how good or how poor the fishing or hunting is or was on some lake or in some woods. They talk about people Jimmy knows. They talk about how many points a buck did or did not have, how many pounds a catfish did or did not weigh. They note local lake temperatures and mention how high the water is. Jimmy usually asks the cameraman to run a video of some

recent fishing or hunting trip.

Then there's the picture board. Fans from across the region send Jimmy hundreds of fuzzy photos of themselves and their catches. In these photos, smiling boys and girls eagerly hold up strings of fish (sometimes really big ones) and Jimmy reads their names and tells viewers at home where and how the fish found themselves hooked. Sportsmen display turkeys, quail, bass, and deer. Jimmy comments on the size or the beauty of the catch. He and his visitors tell old jokes and chortle joyfully.

Jimmy always wears his fishing hat on the air. Even when the president is on. Jimmy may be a little more nervous about the show with Carter, but Jimmy is so laid back in general that it's hard to tell. He shows the former president a fishing lure that looks like a peanut. They talk some about what Carter calls "the great fraternity of people who love the outdoors" good folks, everybody agrees. Thirty minutes pass, and it gets on time to go.

"Ya'll be sure and wear your life jackets," Jimmy tells viewers. "And we'll see you next week." Judging by the last 18 years, he probably will.

The Nashville area is a fisherman's paradise, noted by the more than half a million fishing-hunting permits issued in the area each year. Percy Priest Lake is one favorite fishing spot, with its tranquil waters and abundant game-fish. Photo by Bob Schatz

Locals flock to the lakes for more than fun, however. They go for dinner, too. Tennessee is famous for fishing, and the Nashville area is an angler's dream, as evidenced by the more than 525,000 hunting-fishing permits issued in the area each year. Bass, trout, catfish, crappie, bluegill, and other gamefish abound in local waters, especially in the reservoirs that were created by area dams. Although most popular in the spring and summer, fishing is excellent year-round in the midstate.

For people who like a little more action in their water sports, canoeing is a popular weekend ritual for hundreds of enthusiasts, who head to surrounding rivers for an unpredictable ride on the currents. The Harpeth and the Duck rivers meander through peaceful countryside. But the Buffalo River, one of the few designated wild rivers in the nation, provides white-water challenge.

IN TOWN BUT OUT-OF-DOORS: SUMMERTIME IN THE CITY PARKS

In addition to its intricate web of waterways, Middle Tennessee has parks galore. Metro alone maintains 77, among them Centennial Park, which contains a full-scale replica of the Greek Parthenon. Warner Parks, one of the largest urban park complexes in the country, contains 2,600 acres and is listed on the National Register of Historic Places.

Metro parks are open year-round, but summertime is when they really shine. In addition to traditional park activities, summer brings symphony performances on sloping green lawns, fireworks above the Cumberland River, big band concerts under the stars, a Shakespeare in the Park series, and Greek theater on the steps of the Parthenon. More than 100,000 people show up when the Parks and Recreation Department throws its annual Fourth of July bash at Riverfront Park.

Summertime is also when folks gear up for softball and baseball in Music City, and they gear up in a big way. As many as 20,000 people play in city-sponsored softball leagues on more than 150 city-owned ball diamonds. Two special, four-diamond softball complexes are used for regional tournaments in different quadrants of the city.

Nashville's Sounds Baseball Club drew its five-millionth fan to Greer Stadium in 1988. That's strong local support for Nashville's AAA team, but it hasn't always come easily.

There were plenty of hoots 10 years ago when Larry Schmittou, an ebullient ex-baseball coach from Vanderbilt University, put together a group of investors and fielded the Sounds, the city's first minor-league baseball team in a decade and a half. People remembered rows of empty seats before the Nashville Vols folded 15 years before. Nobody knew whether Nashville was ready for an AAA team. Larry Schmittou didn't know either, but he was certainly willing to try.

The ex-coach and his associates leased land from the city and built a pleasant little ballpark on some softball fields down by the railroad tracks,

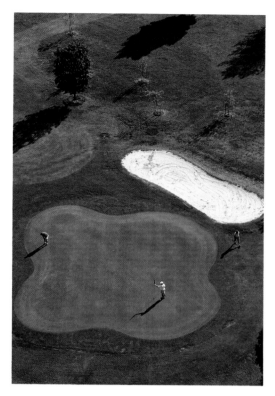

League's Pitcher of the Year award.

Today the Sounds are affiliated with the Cincinnati Reds and play in the American Association. Larry Schmittou remains president and general manager of the ball club, and most importantly, the fans are still coming.

If your yen doesn't run to bats and gloves, there's always the city's new $12-million Centennial Sports Complex, where you can swim, skate, and play tennis. There are nearly 180 city-owned tennis courts scattered throughout Nashville, but the 18 courts in the new complex may prove most important because they undoubtedly will attract increased tournament play. The complex also includes an ice rink (where the Nashville Knights ice hockey team practices) and a competition-size swimming pool.

TEEING OFF

Nashville's public sports facilities are complemented by a variety of private country clubs and athletic centers offering both social and recreational opportunities. There's everything from six neighborhood branches of the YMCA, where you can lift weights by day and take law classes by night, to the exclusive Belle Meade Country Club.

The Nashville climate and topography being what they are, golf is extremely popular here. Seven city parks

paying for the stadium out of their own pockets. Then Schmittou started promoting. And *promoting*. There were bat nights, helmet nights, Schmittou stumpers, fireworks, and performances by the Soundettes. Games took on the aura of a three-ring circus, and fans began to come. And *come*.

Over the next six years, the Sounds averaged 511,000 fans per season—more than any other minor-league baseball team in the nation. Slowly the fledgling club climbed into the black, and management plowed profits back into the park. There were new grandstands, new boxes, new enclosed boxes, a new roof, and sky boxes.

And as the park got better, so did the action on the field. The Sounds fielded championship clubs in 1979 and 1982. They won the 2nd Half Western Division title within the Southern League for six years running, in three of those years going on to win the Western Division crown. During the past decade, more than 100 players have left the Sounds to move into the major leagues, and for six consecutive years the club claimed the Southern

Getting Even Better: The Nashville Aquatic Club

John Morse, head coach at the Nashville Aquatic Club, is happier these days. He has a brand new 50-meter pool, courtesy of the city. For the first time ever, Nashville's national champions will be competing and setting records at home, not just on the road.

A relatively new club, the Nashville Aquatic Club was formed in 1975 through the merger of several small groups into one large, competitive program. Things got off to a great start. Nashville burst onto the national scene in 1978 when local kids Tracy Caulkins, Joan Pennington, and Nick Nevid, among others, became world champions.

Photos by Bob Schatz

A decade later Caulkins remains the grande dame of American female swimmers, having won more events than any swimmer in American history. Heralded as the nation's outstanding athlete by *Sports Illustrated*, Caulkins was at her peak for the 1980 Olympic games, but was denied the chance for gold because of the boycott. However, she continued to train. Four years later, she was older and not as quick in the starts; but she was stronger, wiser, and more efficient. She came home with Olympic gold.

Local kids aren't the only Olympic winners who have trained in Nashville. They've come from Canada, Japan, Australia, and Argentina to develop their skills under Nashville Aquatic Club coaches. The early 1980s saw plenty of record-holders, like local swimmers Patty King and Julie Garde, and in recent years, Nashville native Charles Morton has made an international name for himself.

All the same, as of 1988 or so, some of the program's original energy had sagged—because club facilities were no longer adequate. Morse and the club staff were hard-pressed to maintain an internationally competitive program in their decade-old training pool.

Fortunately, the city goverment and HCA came up with a plan. The city's Centennial Park lies next to HCA corporate headquarters. HCA had land that Centennial could use, and Centennial had land that HCA wanted. They swapped, and HCA threw in $12 million to build a new sports center—the Centennial Sports Complex, a combination ice rink, tennis facility, and aquatic center with an auditorium.

The deal made an Olympic-size pool available to the public year-round and created a competitive facility for Nashville's swim teams as well. Nashville's first national competition was held in 1989. And all that makes coach John Morse a whole lot happier. "We're gonna grow," he predicts. "Things are getting even better."

and most private clubs sport busy golf courses, some of them outstanding. Richland Country Club boasts a Jack Nicklaus-designed 18-holer, and Opryland USA, already known for its amusement park and hotel, has purchased property for a course to be designed by Larry Nelson. Noted golf course architect Pete Dye designed the Harpeth Golf Club course west of town, and Nashville's Hermitage Golf Course hosts the annual Sara Lee Classic for women.

STATE AND FEDERAL PARKS

Four state parks are within an easy hour's drive of Nashville, including 6,000-acre Edgar Evins Park, where campsites overlook Center Hill Lake. Farther away, but still within easy driving distance, is Fall Creek Falls, where you can fish from the porches of comfortable cabins. At Reelfoot Lake there's even an earthquake simulation room, which gives you a feel of what it was

like during the 1811 quake that made the Mississippi run backward to create the lake. Bald eagle watching is a favorite activity at this park, and wooden walkways over the swamp provide an easy (and safe!) method of observing swamp creatures below. South Cumberland Park is for the hiker who likes to rough it, and the beautiful, 11,400-acre Savage Gulf Natural Area contains 85 miles of backcountry trails.

COLLEGIATE ENERGY

At Vanderbilt University, fans pour into Dudley Stadium for Southeastern Conference (SEC) football with the Commodores. The stands seat 41,000, and they fill to capacity on Saturday afternoons when the team is at home facing opponents like the Georgia Bulldogs or archrivals University of Tennessee Volunteers. There's football action at Tennessee State University as well, where the late coach John Merritt built the Tigers into a traditional

At Fall Creek Falls, just a little more than two hours by car from Nashville, refreshing water from the Cascades cools down a hot summer day. Photo by Bob Schatz

81

ABOVE: The Vanderbilt University Commodores clash with the University of Texas Longhorns during an action-packed basketball season. Photo by Bob Schatz

OPPOSITE: From an exhilarating hot-air balloon excursion to an autumnal ride along a quiet bridle path, Nashville offers a wide array of recreational fun. Photo by Rudy Sanders

RIGHT: The game was tied with just minutes to play as Vanderbilt University battled against the University of Kentucky Wildcats. Photo by Bob Schatz

Tigerbelle superstar, became the only track performer ever to win three gold medals in one Olympiad. All told, the Tigerbelles have won 20 Olympic medals since 1950; before his recent retirement, Temple guided nearly 40 women to Olympic competition.

When fall turns into winter, Nashville's big sports news is basketball. The program at Vanderbilt University has produced pro players such as Charles Davis, Will Purdue, and 1984 Olympic team captain Jeff Turner. David Lipscomb College consistently produces championship basketball teams and, as spring approaches, fields outstanding baseball teams as well. Longtime baseball coach and local legend, Ken Dugan, has led David Lipscomb baseball to multiple NAIA championships.

And that's just the beginning. Across the city, Nashville's colleges and universities field outstanding teams, making for lively spectating. At the same time, the area's natural resources provide a challenging and beautiful tableau for warm-weather recreational and athletic endeavors of every description. It all adds up to make Middle Tennessee a good place to play—any time of year.

powerhouse, producing pro players like Ed "Too Tall" Jones of Dallas Cowboys fame.

No doubt the most visible TSU alumni are in the National Football League, but perhaps the most remarkable have been those who have run as Tigerbelles, members of TSU's phenomenal women's track team. Legendary coach Ed Temple guided more than his share of Olympic winners during his 35 years at TSU. The Nashvillian was head coach of the women's Olympic track team in 1960, the year Wilma Rudolph,

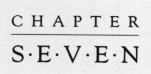
Nashville's Business
is Diversity

"**A** lot of small companies make up the Nashville business community. We don't have one giant company or industry that rules the economic roost," states Don Belcher, director of research for the Nashville Area Chamber of Commerce. "The city's fortunes don't rise or fall on the strength of a single industry."

Nashville businesses are an eclectic mix. In recent years established businesses such as Nissan, Bridgestone, Citicorp Insurance Services, CNA Insurance, General Motors, and Northern Telecom have all relocated to Nashville. In addition, more than 250 new businesses have relocated to the area since 1980. Since 1985 over 40 national and international firms have located all or part of their headquarters here. These companies joined a host of homegrown firms that have risen to national and international prominence: Hospital Corporation of America (hospital management), Opryland USA (entertainment), Genesco (family shoes), Service Merchandise (catalog showrooms), Werthan Industries (bags), and Shoney's Restaurants.

Add to that list music, film, printing and publishing, advertising, medical research, and banking, and the breadth of Nashville's economy is apparent. "What's impressive about Nashville is how so many unrelated industries, like manufacturing and entertainment, do so

well here," comments Jeff Wilson, editor of the *Nashville Business Journal.*

STEADY GROWTH
Nashville is not "a get rich quick town," admits Belcher, "but it is a good place to do business for the long haul." From Nashville's earliest years the business community has displayed a remarkable ability to adapt and expand in response to new opportunities. In the nineteenth century Nashville emerged as a transportation center for rail and barge traffic linking the North, Midwest, and South. A strong financial base stimulated manufacturing, commerce, banking and insurance—historically the mainstays of the local economy. A Saturday night radio show, the Grand Ole Opry, spawned the massive music industry that put Nashville on the map. Local medical schools served as the springboards for the area's important healthcare industry. In recent years support services for Middle Tennessee's burgeoning automotive industry

Nashville is a city of thriving economic opportunity, refreshing spirit, and community pride. Photo by Matt Bradley

have developed around the nearby Nissan plant. More are expected as General Motors nears completion of its massive Saturn complex, located 30 miles south of Nashville in Spring Hill, Tennessee.

This diversity helps make the city recession-proof, effectively insulating it from the travails of the national economy. During the 1981-1982 recession, when the national unemployment rate hovered at 9.7 percent, Nashville's was only 7 percent.

Over the past decade Nashville's unemployment rate has averaged 6 percent versus 9 percent for the nation.

Don Belcher adds that throughout the 1980s Nashville gained over 120,000 new jobs—a 3.3 percent annual increase. That compares with a 1.7 percent annual increase nationally. "We average 1,000 new jobs each month," states Belcher. The majority of these new jobs, over 80 percent, are created by existing local businesses.

Local labor isn't a lopsided picture dominated by one skill. Around 19 percent of the workers hold manufacturing jobs, 24 percent work in the trades, 25 percent provide services, and 14 percent are government employees.

LOCAL INITIATIVE

Nashville is a strong entrepreneurial community. Each year over 1,400 new companies open their doors for business here. *Inc.* magazine lists the city as the seventh hottest spot for business start-ups in the country. More importantly, the magazine ranks the city in third place for the success and growth of those start-ups.

In the tradition of American Express, Kentucky Fried Chicken, and Maxwell House Coffee, which all started here, many new Nashville companies are making their marks on the world: Telco Research (telecommunications), Arnet Corporation (computer production), FISI (bank services), and Advanced Voice Technology (voice mail). "These companies aren't high-profile in the community like HCA," says the *Nashville Business Journal*'s Wilson, "but they're making enormous impacts in their fields at national and international levels. They're typical of the mix of companies that start up and grow here."

Much of Nashville's economic development efforts are focused on helping local businesses expand. "Local businesses are the backbone of our economy," says Howard "Butch" Eley, who heads the mayor's new Office for Economic Development. "My job is to help them grow."

SUCCESSFUL RECRUITING

The city government works closely with the Nashville Chamber of Commerce and Tennessee's Department of Economic and Community Development in recruiting firms to the Nashville area. Over the past 10 years the area has enjoyed incredible success. "The Nissan and Saturn plants were two of the most sought after projects in the country," says Mayor Bill Boner. "They both picked Middle Tennessee."

Nissan and Saturn have invested over three billion in their plants. Nissan began making light trucks in nearby Smyrna in the early 1980s and quickly added car production.

The 400-million-square-foot Saturn plant opened in 1990 with 3,000 workers. That number will increase to over 6,000 when the plant reaches full production later in the decade. Another 16,000 jobs may be created in spinoff industries. Already several support companies have relocated to the Middle Tennessee area.

An $11.5-million subassembly and warehousing plant for AP Technoglass employs over 100 workers. Johnson

Controls, which supplies seats for Saturn, has opened a $28-million production facility. Bridgestone USA Inc. has relocated its corporate headquarters and tire sales division here. Bendix-Jidosha Kiki Corporation is building an $8-million automotive-parts plant in the area. These new companies join the local Ford Glass Plant, which has operated in Nashville for over 20 years and is a major producer of automobile glass in the world.

Companies are attracted to the Nashville area for a lot of different reasons according to Fred Harris, economic

LEFT: Dalcon Computers is one of many full-service computer companies serving the Nashville area. Offering complete systems packages, which include software, hardware, training, and support, Dalcon is just one of Nashville's growing businesses; some 2,000 small companies are located in this healthy and prosperous business community. Photo by Rudy Sanders

development director for the chamber of commerce. Harris Graphics needed a central location to serve its existing markets. For its national financial and data processing center, Deloitte Haskins + Sells, a large accounting firm, needed skilled computer workers. Western Star Trucks was looking for comfortable, affordable housing for its employees. American Airlines established its southeastern hub operation in Nashville because of the area's market potential, diverse economy, and the commitment of the community to continued growth.

ON THE WAY TO EVERYWHERE
The American Airlines hub operation, housed in the city's new airport, is the crown jewel of Nashville's transportation system. Figures supplied by the Metro Nashville Airport Authority show that since the hub opened in 1986, air traffic here has increased from 4.5 million travelers annually to almost 7 million. American operates 120 flights a day through Nashville. Plans call for up to 200 daily flights when the hub is fully established. In all, 11 airlines operate over 400 daily flights in and out of Nashville.

Nashville's central location—the city sits within 500 miles of 76 percent of the U.S. population—and extensive transportation system includes rail, highways, and waterways, which attract many companies to the area.

Three major interstates intersect at Nashville, making the city a regional trucking center: I-65 links Nashville to Chicago and Indianapolis in the industrialized Midwest and provides easy access to Birmingham and the port city of New Orleans; I-40 leads to the important Atlantic and Pacific coast markets; I-24 links America's heartland with Atlanta and Florida.

In just 10 years the number of trucking companies operating here has increased from 40 to over 200. The city is served by a major UPS facility. Almost 24,000 people make their living in trucking and its related industries in Nashville.

The city emerged from the Civil War as the center of the South's first modern rail system, the Louisville and Nashville Railroad. Today, as part of CSX Transportation, Nashville's rail system links local shippers directly to important markets like Cincinnati, Chicago, Memphis, Atlanta, and New Orleans. The local operation is one of seven freight-car classification yards in the CSX System and is the major hump yard for the southeast. The Radnor Yard piggyback-rail terminal is one of the most modern in the country and provides delivery to Nissan, the Ford Glass Plant, Service Merchandise, and Bridgestone among others. The 82-acre terminal features two 50-car loading tracks and parking for over 1,000 heavy trailers. It handles over 6,600 trailers each month.

The Cumberland River snakes its way through the heart of Nashville and ties the city to points on the Mississippi and Ohio rivers and the 234-mile Tennessee-Tombigbee Waterway, a $2-billion barge canal to the Gulf of Mexico. Five common carriers and three public port facilities on the Cumberland serve the chemical, grain, oil, and mining industries. A significant portion of the shipments originating in Nashville is bound for overseas destinations through Gulf ports. American Shipbuilders operates a barge construction facility on the river just across from downtown.

A DECADE OF CHANGE

Over the last 10 years, the face of Nashville business has changed considerably.

The city entered the 1980s as a homegrown town where nearly all of its large, successful businesses—from insurance to music to television—were locally owned and operated. The Nashville of the 1990s is a very different town. Today many of the city's business institutions, including the venerable Grand

Printing and publishing play a major role in Nashville's commercial sector, employing a workforce of more than 13,000 people. Pictured here is Ambrose Printing, just one of the area's many printing companies. Photo by Rudy Sanders

Ole Opry, are owned by out-of-state companies.

Perhaps nowhere has the change been greater than in the city's banking industry. The Nashville banking community is quickly making its mark on the region. Since 1985 three of the city's six major banks have been acquired by super regional banks. The takeovers of Nashville City Bank, Commerce Union, and Third National have provided access to this market by Dominion Bankshares, Sovran Financial Corporation, and SunTrust Banks Inc. Early in 1988, First Union Corporation, a large Charlotte, North Carolina-based bankholding company, opened its Tennessee headquarters in Nashville, and with assets over $29 billion, quickly established itself as the city's largest bank.

TUNES AND TOURISTS

Nashville has long been the country music capital, but in recent years a wide range of pop, gospel, and rock musicians have also turned to Nashville to create their special sound. Paul McCartney, Sandi Patty, Amy Grant, Neil Young, and Bob Dylan are among the artists who have recorded here. All the major record labels, including CBS, RCA, and Warner Brothers, have offices on Music Row. Recently three new record labels opened for business here—Arista, Atlantic, and Curb. Including spinoff industries like booking agencies, music publishing companies, promotional firms, trade publications, as well as performance rights associations such as BMI and ASCAP, the music industry pumps millions of dollars annually into the local economy.

The Nashville music scene has expanded to include music-related advertising firms, especially jingle houses, and music video production. The city is home to two cable networks devoted exclusively to country music, the Nashville Network, which reaches over 45 million homes, and the new CMTV.

The city's strong tourism industry is tied to the success of the music industry. Each year more than 7 million tourists

spend vacation time in the Music City. They visit Opryland, Music Row, take bus tours of the city, and see the homes of the stars. Along the way they spend over $900 million in hotels, restaurants, and shops.

As Nashville businesses prepare for the coming century, they will build on more than 100 years of growth and prosperity. The city is blessed with an abundance of enterprises that have successfully grown up here and stayed. No, Nashville isn't a gold-rush town with quick fortunes waiting to be made, but the businesses here are doing very well. Very well indeed.

A popular destination for Nashville tourists is the remarkable Opryland Hotel that features the Cascades— a breathtaking indoor water garden with a six-story waterfall, synchronized fountains, lights, and music. Photo by Matt Bradley

Nashville's Future

For years Nashville was simply a southern town whose public heart and soul were linked to the vagaries of country music.

Journalists and tourists descended upon the city in search of the quintessential Nashville experience: a meal at a popular "meat and three," a cold beer at Tootsie's Orchid Lounge, and maybe, just maybe, a look at a real, live country music star. Country music. That was Nashville's public persona.

Today although country music plays an important role, Nashville's hospitals, colleges, and banks are nearly as well known as its entertainers. Nashville's public image is that of an up and coming southern town, potentially one of the most influential cities of the coming new century.

Change is often subtle, making it difficult to pinpoint the precise moment when the forces of influence shift. In Nashville's case though, that moment is perceptibly linked to January 1975 when *Harper's* magazine identified the city as one of the most livable places in America. Nashville was quickly cast in a new national light. No longer just Music City U.S.A., the city was crowned with new accolades. It was touted as one of the top ten growth markets of the 1970s, listed among the top three test markets in the country, and identified as the city with the third healthiest economy in the nation. Country music became almost secondary to Nashville's livability, its colleges and universities, and its diverse economy.

Suddenly, everyone was interested in Nashville. The city's advantages were obvious: central location, well-trained and loyal workforce, pleasant climate, and low taxes. Businesses began to relocate here and the population boomed. The opportunities for the city and its people seemed endless and were quickly embraced.

Expansion during the 1970s prepared the city and its residents for the dynamic growth of the 1980s. The city developed into a regional banking power and a center for medical research. It nurtured telecommunications-based high-tech development. With the construction of Nissan and General Motors plants, the city positioned itself as an important developing automotive center. The new airport, and American Airlines' decision to locate a hub operation here, placed the city in the forefront of air travel.

In the face of this rapid change, Nashvillians have managed to preserve their most precious asset—their sense of community. Over half a million people live in the Nashville area and most of them have found this Middle Tennessee city to be their place. There's something wonderfully satisfying about life here. It's just a great place to call home.

In Nashville there's an exciting mixture of Old South and New South. "We're high-tech but with heartland values," is the way Irby Simpkins, publisher and president of *The Nashville Banner*, likes to put it.

Nashville actively competed with other cities for its share of corporate relocations and regional headquarter operations. That effort will continue as Nashville emerges as a leading Southern city in the southeast. Already the city sits within a day's drive of over 50 percent of the American population, and because the U.S. population center is gradually moving westward, Nashville is poised to play an important national role. "We're a natural as a national center," comments Ed Stolman, a noted Nashville investor. "With our interstate system and airline hub we're an easy city to get to so tourism and business can thrive."

In the future Stolman hopes that the city's downtown will be extensively developed. "Any city that's great has a vibrant downtown," he says. He notes that there's a great deal of untapped poten-

tial in downtown Nashville. "The river and some of those grand old buildings have the makings of a terrific downtown. We just need the vision to see the possibilities."

Of growing importance to the continued strength of the city is an organized regional planning effort. For years Nashville was the focus of job growth in Middle Tennessee. The surrounding counties served primarily as bedroom communities. Over the last decade office parks and shopping centers have expanded to these areas. "As growth has spread, Nashville and the adjacent communities have become more interdependent," explains Jeff Browning,

director of the Metropolitan Planning Commission, "and we need to coordinate our planning efforts."

Indeed, managing growth looms as Nashville's biggest challenge in the coming years. As the city charts its course for the future it must do things like expand its infrastructure and continue to improve its schools so that "everyone can benefit from the growth," says Irby Simpkins.

Bertrand Russell has been quoted as saying: "We'll never have a better world until some good people build one." Nashville, Tennessee, has its eye on a better world and it has the good people to build it.

Considered to be one of the most livable places in the nation today, Nashville is poised on the brink of its explosive future, offering a lifestyle that blends the traditional values of the Old South with the contemporary influences of the modern world. Photo by Bob Schatz

P A R T

2

NASHVILLE'S

ENTERPRISES

Networks

N ashville's role as a modern, thriving metropolitan center is made possible by its network
of energy, communications, and transportation
providers.

Photo by Rudy Sanders

Nashville Electric Service

Nashville Electric Service has had a long, proud, and colorful tradition of service since the day it opened its doors, August 16, 1939, with the stated mission of providing low-cost, high-quality electric service to the people of the Nashville area. By the end of that first year, NES was serving 52,063 customers.

Today the only things that have not changed are the company's 700-square-mile service area and the promise of high-quality, low-cost service.

NES customers now use more than 9 billion kilowatt-hours per year, though they pay less per kilowatt than most other Americans, fulfilling the firm's original commitment to safely provide reliable electric power at the lowest-possible cost. Nashville consumers pay only $56 per thousand kilowatts, compared to $120 in New York City, $95 in Detroit, and $65 in Atlanta.

Nashville Electric Service management has made a strong commitment to the development of its commercial and industrial division, which assists businesses that are considering relocation or expansion in the Mid-South area. The company provides confidential information to industrial and commercial executives, reducing their time and cost in making a site selection. NES presently offers a free site-selection handbook that provides valuable information about industrial and commercial sites.

Over the years NES has been just as effective promoting energy conservation, working

closely with area planners, developers, architects, engineers, and contractors. One of the best examples is the Energy Saver Home (ESH) program, which began in 1985. Today 33 Middle Tennessee contractors are building new homes to the demanding ESH specifications. The firm also provides free energy studies of existing homes, as well as offering low-interest, long-term loans for the installation of energy-efficient heating pumps.

In 1989 NES' counselors for senior-citizen customers made more than 13,000 calls, answering questions and special requests, including $357,000 in financial assistance from government and private agencies. Project HELP, a program designed to provide financial assistance to elderly and disabled people who are having problems paying their energy bills, provided another $185,000, contributed by NES customers who volunteered to add one dollar to their monthly electric bills.

National Electric Service's energetic efforts to improve and expand service while keeping rates low have been a formidable challenge. And considering the energy-cost differences between Nashville and most other American cities, the challenge has clearly been met.

Metropolitan Nashville Airport Authority

When Berry Field opened in 1936 as Nashville's municipal airport, the original 340-acre tract was home to only a three-story terminal building with a control tower and one paved runway. The airfield was named Berry Field in honor of Colonel Harry S. Berry, Tennessee's administrator of the federal government's Works Progress Administration.

The airport became the base for military

operations during World War II. In 1946 the military returned an improved airport comprising 1,500 acres to the City of Nashville. A new Air Traffic Control Tower was constructed in 1954, and by 1958 the city aviation department had begun to plan for a new terminal. Three years later a new 145,900-

square-foot terminal opened to the public. The existing runway was extended by 600 feet in 1963, and construction on a new runway was begun.

Nashville's newly formed metropolitan government, seeing the growing importance of air travel to the city and pursuant to 1969 state-enabling legislation, formed, in 1970, the Metropolitan Nashville Airport Authority (MNAA)—a self-supporting public corporation that would manage and operate a system of regional airports. The MNAA was mandated to operate without the benefit of local sales and property tax dollars and to manage Nashville's airport to the same economic standards as those of private business. In 1973 MNAA completed a master plan for long-term growth of the airport calling for a new terminal and runway.

In 1980 the MNAA updated the 1973 master plan and began an environmental assessment for the new terminal. By 1984 site preparation for the new terminal had begun. The following year American Airlines announced its selection of Nashville as the location for a major north-south hub. Soon after, the number of signatory airlines increased as additional airlines demonstrated their confidence in Nashville's market potential by signing long-term lease agreements. At the same time, confidence in the growth of the city of Nashville reached new heights, as businesspeople nationwide acknowledged the large contribution the expanded airport would make to Nashville's economy.

In 1986 American began its hub operation. On September 14, 1987, service began from the $110-million, 750,000-square-foot, state-of-the-art airport terminal complex. Just over a year later Nashville's scheduled airlines were providing service to more than 100 cities. By November 1989 the Metropolitan Nashville Airport Authority had fulfilled the final major demand of its 1973 master plan, when a new $78.5-million parallel runway, which allows simultaneous takeoffs and landings, opened at Nashville International Airport.

Nashville's 340-acre Berry Field served the city's first wings. Courtesy, postcard collection of Walter and Brenda Jowers

A Boeing 727 takes wing from Nashville International Airport's 750,000-square-foot, state-of-the-art terminal.

WLAC AM/FM

ABOVE: WLAC-AM, with Dominion Bank, awarded the New Business Grant to Bill Kearns, inventor, entrepreneur, and president of Elecom, Inc. Kearns (center) is shown congratulated by senior management of WLAC and Dominion Bank.

RIGHT: 106 FM morning personalities Terry Hopkins and Phil Valentine display a pair of new friends for a listener whose own dog had recently been killed.

From its beginnings in 1923 as shortwave radio station WDAD to the 50,000-watt community leader known today as WLAC-AM and 100,000-watt WLAC-FM/106, the history of WLAC radio is one of trend-setting firsts, colorful air personalities, forward-looking management, and consistent growth.

In 1926 the Life and Casualty Insurance Company (L&C) bought shortwave station WDAD. The station was so named because of its location over Dad's Auto Store. The new owners renamed the 5,000-watt station WLAC, and the fledgling station immediately began making radio history. As one of the

original affiliates of the CBS Radio Network, WLAC produced popular daytime shows that featured a staff orchestra and singers. The station originated a Saturday-morning program called "The Garden Gate" that ran for 20 years.

Since the 1930s WLAC has had a history of colorful and innovative announcers. An early air personality was Tim Sanders, who was famous for a program called "The Air Traveler Airs His Views," in which he interviewed the rich, famous, and celebrated as their planes from New York to Los Angeles stopped for refueling at the Sky Harbor Airport near Nashville.

Also renowned was Herman Grizzard, the "Old Colonel." As the sports announcer for the Nashville Volunteers minor-league baseball team, Grizzard perfected the art of calling the game's play-by-play commentary, even though he was working from a Morse code transcription of the game. Grizzard went on to become the teacher of such illustrious baseball announcers as Mel Allen, Red Barber, and even Ronald "Dutch" Reagan.

In the mid-1950s WLAC became the first power station in the country to play music by black artists as part of a popular music format. Popular WLAC air personalities of this period were Gene Nobles, Hoss Allen, and John R., who broadcast music and voices through the clear channel of the night, selling everything from baby chicks to hair-care products. During this period performers such as Otis Redding would appear at the station with a new recording in hand, offering it to the WLAC disc jockeys as an exclusive. WLAC's broadcasts in the 1950s and 1960s greatly helped to establish rhythm and blues as a part of popular culture.

Today WLAC AM/FM, located at 10 Music Circle East, employs more than 50 full-time and 15 part-time staffers. As part of Fairmont Communications, the stations have established a strong position in the radio marketplace. As it has throughout its history, WLAC continues to meet market needs with spirit and dynamic energy—from rock to adult-contemporary music, as the flagship station of the Vanderbilt University sports teams, and, through news and community action, as the CBS affiliate.

The Tennessean

Some of the brightest names in American journalism have been associated with *The Tennessean* during the 178 years of its existence. Founded in 1812 as *The Nashville Whig,* today's newspaper is the result of 15 consolidations over the years. Throughout most of those years, particularly in the twentieth century, *The Tennessean* has been a strong editorial voice in its region.

One of the early ancestor papers, *The Nashville Union,* was organized as the home political organ of Andrew Jackson. For one year after the Civil War, another ancestor paper was partially owned by Henry Watterson, who left Nashville to take over the *Louisville Journal.*

The modern history of the newspaper began on May 12, 1907, when Colonel Luke Lea, a

LEFT: *Chairman, publisher, and chief executive officer John Seigenthaler.*

FAR LEFT: *Editor Frank Sutherland.*

colorful politician, organized a new paper, *The Nashville Tennessean.* Grantland Rice was the paper's first sports editor. Within a few years Lea's new paper absorbed a leading competitor, *The Union and American.*

Morning, evening, and Sunday editions were published by Colonel Lea until March 1933, when, as a result of the Depression, the papers were placed under a federal receiver, Lit J. Pardue, an attorney and former newsman. Under Pardue's guidance the papers grew, and in 1937 Silliman Evans, Sr., a Texan, bought *The Tennessean* at public auction.

Evans revitalized *The Tennessean* and remained its chief executive until his death in 1955. His son Silliman Jr. was publisher until 1961, when he died, and then the leadership passed to his younger brother, Amon Carter Evans. In 1979 the Evans family sold *The Tennessean* to the Gannett Company.

In 1962 John Seigenthaler was named editor of *The Tennessean,* and later he added the additional titles of chairman and publisher. In 1982 Seigenthaler assumed the additional role of first editorial director of *USA Today,* the nation's first national, daily, general-interest newspaper. In 1989 Frank Sutherland was named editor.

In mid-1990 *The Tennessean,* which handles advertising, circulation, and printing for itself and Nashville's afternoon paper, the *Nashville Banner,* will begin printing the two papers on new state-of-the-art computer-controlled offset presses.

Over the past 40 years, well-known journalists who have worked at *The Tennessean* include David Halberstam, Tom Wicker, Will Grimsley, Ed Clark, Wallace Westfeldt, Patrick Anderson, Creed Black, Jesse Hill Ford, Bill Kovach, Fred Graham, Elizabeth Spencer, James Squires, Wendell Rawls, Jr., Reginald Stuart, and U.S. Senator Albert Gore, Jr.

The Nashville Banner

The *Nashville Banner* was founded on April 10, 1876, by veteran Nashville newspapermen and printers. They were William E. Eastman, who served as the newspaper's first president; John J. Carter, secretary/treasurer and sports writer; Thomas Atchison, editor-in-chief; R.J.G. Miller, city editor; Cicero Bledsoe, reporter; Homer Carothers, composing-room foreman; and Pleasant J. Wright, who also worked in the composing room.

In 1893 Major Edward Bushrod Stahlman became sole owner of the *Banner*. Stahlman headed the newspaper until his death in 1930, at which time leadership passed to his grandson James Geddes "Jimmy" Stahlman. In June 1972 Jimmy Stahlman retired from his 60-year newspaper career, six months after selling the publication to Gannett Co., Inc., a national newspaper group.

The *Banner* was sold in 1979 to Music City Media, Inc. Today it is the only locally owned daily newspaper in Nashville. Brownlee O. Currey, Jr., is chairman of the board, and Irby Simpkins, Jr., is the newspaper's president and publisher.

The newspaper is a perennial leader in community affairs, sponsoring programs such as its community luncheons, in which members of the public are invited to discuss with the publisher topics of interest ranging from education and literacy to crime and federal courts. The top 100 privately held companies in the mid-state region are recognized yearly at the Greater Nashville 100 Reception sponsored by the *Banner*. Local business leaders who have been featured in the newspaper are honored each year at the *Banner*'s Business Profiles Reception.

The *Banner* has helped hundreds of mid-state students through its educational programs, such as the Bootstraps Awards, honoring high school seniors who have overcome obstacles or problems in life. Each year five of these students receive college tuition scholarships. Minority students are offered six weeks of on-the-job journalism training,

Publisher Irby Simpkins, Jr. (seated), and editor Eddie Jones.

The Nashville Banner *delivers news of the world as well as news of Nashville to the city.*

and the most outstanding student is presented a scholarship through the *Banner*'s Minority Youth Program. Other educational opportunities sponsored by the *Banner* include a spelling bee, writing competition, and art exhibition.

In mid-1990 *Banner* printing operations will be moved to a new four-story press building that is being built next door to the *Banner*'s offices, on the site of the former Sealtest Ice Cream building. The new computer-controlled offset presses will produce sharper print with less ink rub-off, improving the overall quality of the newspaper.

The work of the *Nashville Banner* and its staffers has been recognized with awards such as the Ida B. Wells Award for distinguished service in the field of race relations, the Tennessee Press Association Award, the University of Missouri Photo Competition, and the Tennessee Associated Press Managing Editors Contest.

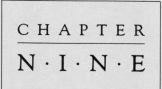

Industry

P roducing and distributing goods and foodstuff for consumers, industrial firms provide employment for many Nashville area residents.

Photo by Bob Schatz

Textron Aerostructures

Textron Aerostructures, a subsidiary of Textron Inc., has earned worldwide respect for its innovation and competitive presence in aerospace technology and manufacturing. Using a facility that encompasses more than 2 million square feet, the division produces a wide variety of airframe structures using both advanced metals and lighter composite materials.

Just as Aerostructures is firmly established in the aerospace industry, the company is firmly established in Nashville and continues a tradition that was born more than 50 years ago.

Founded in Nashville—a city once known as the aviation capital of the Southeast—by the Aviation Manufacturing Corporation, the division's contributions to aviation began in 1939, when the Stinson Airplane Company built Stinson personal airplanes. The Aviation Manufacturing Corporation bought the facility in 1940, renamed it Vultee Aircraft Company, and began building Vultee Vengeance dive-bombers during World War II for the British Royal Air Force. In 1945 the division served as a second-source production site for Lockheed's famed P-38 Lightning aircraft.

The company's presence as one of the world's leading aerospace subcontractors will

Aerostructures has long been recognized for its valuable contributions to the defense of the United States. Since 1953 the company has built empennages (tail sections) for Lockheed's C-130 Hercules aircraft. Users of the C-130 include the Tennessee Air National Guard. Aerostructures built wings for Lockheed's C-141 Starlifter, and for the massive C-5 Galaxy transport—the free world's largest aircraft and a forerunner of commercial aviation's wide-bodied transports. Wings for Lockheed's L-1011 Tristar were also built in Nashville.

Aerostructures was a major supplier to Rockwell International when it built wings for the U.S. Air Force's B-1B long-range combat aircraft. Contributions to space exploration range from components for lunar modules to the space shuttle program.

As the aerospace industry changes and less emphasis is placed on military products, Aerostructures has changed its focus to commercial aviation. The firm's partnership with Airbus Industrie also reflects the trend toward internationalization, as contractors and countries search worldwide for companies that can provide the highest-quality, lowest-cost aerospace products.

Many of the innovative processes and capabilities sought by prime contractors have been pioneered in Nashville. Recently, Aerostructures developed and implemented a technique of using a pressurized oven to facilitate the manufacture of large structural panels of varying thicknesses and configurations. The process provides repeatable, high-quality results at much higher production rates than had formerly been achieved. Aerostructures is currently applying this procedure to produce test panels for the nation's space program. The test

BELOW: Aerostructures' engineers perfected the technique of using a pressurized oven to facilitate the manufacture of structural aerospace components on a large-scale basis.

extend into the twenty-first century as Aerostructures participates as a risk-sharing partner with Airbus Industrie, a European aircraft manufacturing consortium that includes France, Germany, Spain, and the United Kingdom. Aerostructures will also continue programs with Gulfstream Aerospace on the Gulfstream IV business jet and with Hercules Aerospace on the Titan IV expendable launch vehicle, as well as a manufacturing program for the British Aerospace BAe 146 aircraft.

RIGHT: Computer-aided design and manufacturing of aerospace hardware and the tools to construct it are making it possible to build aircraft that will fly farther and longer.

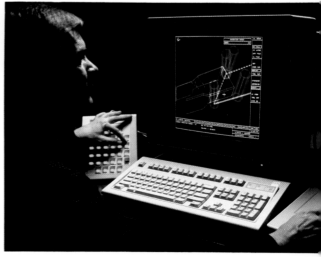

panels resulting from the process mean that it is possible to produce components having the same strength but with less weight, thereby al-lowing space vehicles to carry greater payloads.

Aerostructures' ongoing research and devel-opment has produced new high-speed tech-niques utilized in the construction of nonmetallic aerospace structures known as com-posites. The Nashville operation is a leader in the development of "smart skins"—composite materials that use fiber optics to produce aircraft structures that "talk" and alert pilots and main-tenance personnel to structural conditions.

To ensure it manufactures products of the highest quality, Aerostructures employs an ef-fective and progressive inspection philosophy, from receipt of the materials to delivery of the final product. Computer-aided design and man-ufacturing systems, coupled with coordinate-measuring machines, enhance the ability to rapidly inspect structures during production and final assembly.

Aerostructures' employees have high tech-nical capability, receiving ongoing classroom and on-the-job skills training. Professional de-velopment is encouraged through two on-site college degree programs as well as financial as-sistance toward the pursuit of degrees in the university setting.

Employee energy and direction are focused on maximizing creativity and performance. A

progressive labor/management relationship has resulted in mutual respect, open communica-tion, and team building. The company has compiled an impressive track record through dedication to serving the customer.

Textron Aerostructures' vision, "To be the best supplier of aerospace products in the world," reflects a dedication and spirit that will guide humanity into the ultimate frontier. The company is proud to have shared this spirit for more than 50 years with the people of Nashville.

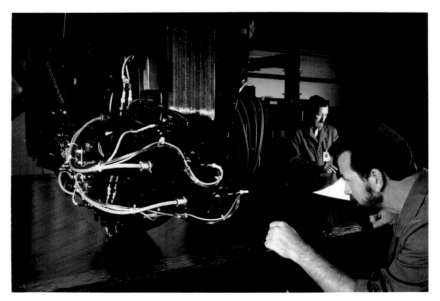

Robert Orr/SYSCO

Nashville, near the geographical center of the state of Tennessee, is renowned for its music and publishing businesses. Yet Nashville is also the home of Robert Orr/SYSCO, the largest full-line food-services distributor in the southeastern United States.

In a growing, diversified metropolis the size of Nashville, food equally as diversified yet necessarily pleasing to the palate is essential. Robert Orr/SYSCO, one of the largest companies in the number-one worldwide SYSCO food-services network, is ready and able to meet this demand. The certified Master Chef on staff only tells part of the story.

It all began just prior to the War Between the States. In 1859 Robert Orr, Sr.'s, trading post on the Cumberland River began supplying the Indians; later, both Union and Confederate armies received dried rations from Robert Orr. In the 1880s the company began distributing canned products, and in 1920 another growth phase occurred. Walton Cunningham, Sr., purchased the firm from Robert Orr, Jr. Then, in 1928, the company secured Maxwell House

Coffee's equipment when Maxwell House moved from Nashville. From this resulted purchases of South American green coffee beans and the eventual packing of coffee under the renowned Hermitage label.

Frozen-food lines (fruits, vegetables, breads, seafoods, meats, and fowl) emerged in 1957. In 1975 the company moved to its present location at One Hermitage Plaza (previously Centennial Plaza).

A 1972 merger greatly facilitated this move. The Robert Orr Company entered the network of Systems and Services Company (SYSCO). SYSCO emerged when seven independent food distributors joined in 1969 to form a national food-services corporation. Their goal was to offer an integrated system of companies that could better serve the entire American economy. Today this goal is firmly in place with 150 SYSCO representatives who purchase products from prime growing regions in the United States, and more than 50 overseas agents who source and ship fresh produce to the United States throughout the year.

Burton Hummell, president and chief executive officer of Robert Orr/SYSCO.

As part of SYSCO, Robert Orr/SYSCO is the vital link between growers, manufacturers, processors, and customers. The bottom line is that customers receive the widest-possible selection of superior food products. This is significant in the growing trend of meals eaten away from home, plus new demands for healthier foods and fresher, more natural produce.

In addition to the Nashville office, Robert Orr/SYSCO has offices in Knoxville, Tennessee, and Birmingham, Alabama. All three cities also contain a self-service distribution center. In Nashville, consumers can purchase any available item from an individual package of paper napkins for the family

The employees of Robert Orr/SYSCO work together to bring a wide array of high-quality food-service items to a diverse and demanding industry.

reunion to a case of sheet cakes for the company picnic to an individual serving of Chicken Cordon Bleu.

The express transportation network of Robert Orr/SYSCO expands this personalized service to parts of 11 states from its Nashville warehouse, with the majority of products distributed to Tennessee, Alabama, and Kentucky. Fresh produce, fresh meat, fresh seafood, frozen fruits and vegetables, dry groceries, chemicals, paper products, and restaurant supplies are delivered to food-service institutions throughout the area, including hospitals, restaurants, businesses, and schools.

With this quantity of over-the-road travel, a dependable truck fleet is essential. Here, as elsewhere, the firm's innovation shines through: All trucks are inspected closely for operation and safety efficiency, and all repairs and replacements are performed at Robert Orr/SYSCO's on-site vehicle maintenance division. And the company's uniformed service representatives (delivery personnel) undergo a rigorous training program before assuming their duties.

Innovation permeates the organization: The dedication to fresh foods is enhanced by a huge 350,000-square-foot warehouse containing dry, chilled, and frozen (zero to minus-six degrees Fahrenheit) divisions; delivery trucks are similarly compartmentalized. A prototype delivery truck called "The Hummer" (named after Robert Orr/SYSCO president Burton Hummell) is considered innovative in its urban-delivery capabilities. In addition, the hiring and training of specialists is priority in each of the company's specialty departments—fresh produce, seafood, meats, beverages, chemicals, and equipment and supplies—most large enough to succeed independently.

And for coffee clients desiring to serve a full-bodied cup of coffee, the company has the Robert Orr/SYSCO Supreme Coffee Award. This honors those establishments meeting the highest standards in coffee brewing, equipment, and personnel training.

Naturally, the sales force (marketing associates) at Robert Orr/SYSCO is an intimate part of the firm's past and its continuing success. At 150 strong, these men and women can rely not only on each other's professionalism and the company's dedication to excellence but they can also depend on a weekly updated 15,000-plus-item copyrighted product book. This contains much more than representative listings of Robert Orr's warehouse stock of about 17,000 items, since it gives the sales force ready-reference access to facts, figures, and general information.

The product book is unique in the food-services field. A variety of in-house-designed catalogs, brochures, and other valuable sales aids are also available to all marketing associates.

Aside from many of the expected ways a long-established company participates in the Nashville community, Robert Orr/SYSCO regularly leads the way. For example, the firm played a significant part in the famous Rotation Diet. Dr. Martin Katahn, director of the Vanderbilt Weight Management Program and the diet's originator, met Robert Orr's president during one of Katahn's weight-loss programs. Dr. Katahn consequently gained an "inside connection" to the food business. Teamwork between Hummell, Katahn, and others at Robert Orr resulted in Katahn's business choice for his diet's weigh-in centers, Robert Orr designing and printing the Rotation Diet brochures, and the company designing the diet's logo.

The rest continues to be history.

Robert Orr/SYSCO's level of professionalism, plus its dedication to quality in products, services, and people, is the signature of a Nashville and southeastern U.S. business leader growing into the 1990s. Robert Orr/SYSCO—a tradition of excellence through innovation.

Bridgestone

In 1928 Shojiro Ishibashi established at Japan's Nippon Rubber Company a tire division that grew in just three years to be an independent operation—Bridgestone Tire Co., Ltd.—with Ishibashi as chairman and president. The company name came from the English translation of the name Ishibashi, which is "stone bridge." Thinking that no one would buy a tire named Stonebridge, Ishibashi transposed the words and named his tires Bridgestone.

The firm had great early success as the Japanese subsidiaries of Ford, General Motors, and Chrysler all chose Bridgestone tires as standard equipment for their cars in 1932. Bridgestone first diversified in 1935, adding the manufacture of golf balls. Through the years the company has continued to diversify, but tires remain its primary product. At present Bridgestone manufactures more than 8,000 different types of tires, as well as golf balls and clubs, tennis balls and racquets, robot arms, bicycles, car and truck wheels, and automotive seats and seat belts.

Bridgestone Corp. presently holds more than 50 percent of the original equipment automobile tire business in Japan. During the 1980s the company won important prestige tire contracts: the Porsche 959, 944, and 928 models and, most recently, the Ferrari Testarossa model.

Bridgestone first arrived on the U.S. tire market in the form of Bridgestone Tire Company of America. Founded in 1967 in Los Angeles as a sales and marketing arm of Bridgestone Corp., the office had a staff of just seven people.

At the start all Bridgestone passenger and truck tires sold in the United States were imported from Japan. By the early 1970s, though, Bridgestone Tire Company of America had outgrown its original Los Angeles offices and

moved its sales staff of 140 people into a new facility in Torrance, California.

In January 1983 Bridgestone Corp. of Japan purchased a truck- and bus-tire manufacturing plant from Firestone Tire & Rubber Co. Bridgestone Tire Manufacturing (U.S.A.) Inc. was formed to operate the facility, located in La Vergne, Tennessee, making Bridgestone Corp. the first Japanese tire maker with a U.S. tire plant. At the time of the acquisition, 450 of the plant's previous employees had been laid off. The remaining 235 workers were producing just 16,400 truck tires per month.

Bridgestone Tire Mfg. began recalling all laid-off employees, and by the end of 1983 production employment totaled 481 people and monthly production had more than doubled. Production employment in 1987 reached 595 workers, and some 82,175 radial truck tires were being manufactured each month. Since the acquisition of the La Vergne plant, Bridgestone (U.S.A.) has invested more than $450 million to expand and improve the facility.

In March 1988 Bridgestone Corp. acquired the Firestone Tire & Rubber Co., giving Bridgestone a greater foothold in the global tire market. In late 1988 Firestone announced plans to spend some $1.5 billion to expand its tire manufacturing capability. Part of that expenditure allowed the manufacture of Bridgestone-brand passenger tires at two North American Firestone facilities. Today Bridgestone Corp. operates 22 plants worldwide. In addition, Bridgestone Corp. assists in the operation of the 16 Firestone plants, giving Bridgestone Corp. some 38 manufacturing facilities around the globe.

In late 1986 Bridgestone Tire Co. of America and Bridgestone Tire Manufacturing (U.S.A.)

were merged into Bridgestone (U.S.A.) Inc., with separate sales and manufacturing groups.

In May 1988 Bridgestone (U.S.A.) began manufacturing radial passenger tires at the La Vergne facility, becoming the first Japan-based tire company to manufacture such tires in the United States.

Later that year Bridgestone (U.S.A.) Inc. moved its corporate headquarters from Torrance, California, to Nashville.

In February 1989 Bridgestone (U.S.A.) announced plans to build a new $350-million radial truck-tire plant on some 900 acres in Warren County, Tennessee, near McMinnville. Commercial production is slated to begin in early 1991. The Warren County plant will employ 750 workers and will have an initial output of about 2,000 radial truck tires per day. Ultimately the plant will produce some 4,000 truck tires daily.

In August 1989 Bridgestone (U.S.A.) was consolidated with the Firestone Tire & Rubber Co. to form Bridgestone/Firestone Inc., headquartered in Akron, Ohio. The Nashville headquarters became the base of operations for the Bridgestone Division of Bridgestone/Firestone Inc. The Bridgestone Division handles the marketing and sales of Bridgestone-brand tires in the United States.

Since making Middle Tennessee its U.S.

home, the Bridgestone Division has become symbolic of the great success and cooperation that Japanese companies have enjoyed in the state. Few companies better demonstrate the loyalty and harmony that are natural to the Bridgestone family. From American nicknames for its Japanese staff to American schoolchildren who have learned the ancient Japanese art of origami (paper folding), Bridgestone has united two cultures and discovered their common values.

ABOVE: An excellent relationship between Japan and the United States exists at the Nashville Bridgestone plant.

BELOW: Bridgestone Park is the firm's expansive headquarters in Nashville.

Nissan Motor Manufacturing Corporation U.S.A.

Robotic welding is evidence of Nissan's automation technology.

After nearly 40 years of manufacturing high-quality automobiles for a worldwide market, Nissan began to search in the 1970s for a site for its first U.S. manufacturing plant. The company's objective was to build its plant in an area with a talented work force, easy access to suppliers and markets, and a supportive business climate. In October 1980 Nissan announced that the company would establish Nissan Motor Manufacturing Corporation U.S.A. at Smyrna, Tennessee, a few miles southeast of Nashville.

Nissan's single goal was to build the highest-quality vehicles sold in North America. To that end, the company made and maintains a commitment to treat its employees in the Smyrna plant as its most valued resource. Though the Smyrna plant features state-of-the-art technology and automation, Nissan management realizes that it is people who run the machines, solve the problems, and find better ways to build automobiles. The company believes that the most effective way to get people to contribute their knowledge and skills is through participative management, in which each of the more than 3,500 members of the Nissan team offer important input to each of the company's six management levels.

To ensure that the Smyrna work force will continue to build some of the finest vehicles sold on the continent, Nissan conducts extensive training for all employees, both before they come to work and on the job. Using this training, employees function as their own quality inspectors; they are responsible for the results of each task they perform. To give employees immediate feedback on quality, random vehicles are taken off the final production line each day and given exhaustive quality and performance tests. Since the first truck rolled off the Smyrna assembly line on June 16, 1983, the company's light trucks and Sentra passenger cars have consistently achieved high ratings from outside testing organizations and customer satisfaction surveys.

The Smyrna plant currently has a production capacity of 240,000 vehicles per year. The company also produces rear axles for trucks, bumper fascias, and engines for Sentra automobiles. By the summer of 1992 an additional $490 million will be invested in the plant, and 2,000 more jobs will be created with the start-up of production of a new four-door car. The plant's annual production capacity will then be increased to 440,000 vehicles.

Nissan's Smyrna plant makes a significant

Employees in the paint plant put the finishing touch on shiny red Nissan Sentras, soon to please the eyes of American drivers.

contribution to the local, state, and national economy. The plant employs 3,500 people locally and indirectly supports another 10,000 American jobs. More than 40 firms that supply Nissan are located in Tennessee, and 13 of those firms built facilities in the state specifically to supply Nissan.

As a corporate citizen, Nissan Motor Manufacturing Corporation U.S.A. is proud of the support it gives to nonprofit organizations in the community. The company is a major contributor to the local United Way and has twice won the Greater Nashville Area Chamber of Commerce Business in the Arts Award.

The Bailey Company

The Bailey Company was started by James M. Bailey, then a regional salesman for a Memphis firm. In 1949 the organization sold its first Towmotor forklift truck, a machine that was a true innovation of its time, making work that once was laborious and time consuming easier and more efficient. The Caterpillar lift trucks, sold and maintained by The Bailey Company, are the direct descendants of that early machine; these trucks set the current industry standards for high-quality lift trucks.

The firm is still owned by the Bailey family and operated by officers with years of experience in the field. Company president Gordon Morrow has more than 30 years' experience, as does vice-president/product support Charles Bebout. Vice-president/sales Fred Osborne has been with the organization for more than 20 years. Bert Bailey, executive vice-president/treasurer, and Laura Bailey Busby, vice-president/comptroller, each have more than 15 years' experience.

As manufacturing and warehousing in Middle Tennessee continue their rapid growth, The Bailey Company is prepared to meet the rising demand for lift trucks and other equipment. The company is the largest of its kind in Middle Tennessee and one of the top 10 dealers of material-handling equipment in the United States, with sales and maintenance offices in Nashville, Chattanooga, Knoxville, and Johnson City, Tennessee, and Dalton, Georgia, as well as branch offices in Cookeville, Cleveland, and Tullahoma, Tennessee, and Calhoun, Georgia. The Bailey Company employs 24 sales engineers and four engineering specialists in systems and construction. The firm handles not only new and used lift trucks but also conveyor systems, storage systems, mobile cranes, and industrial boom lifts.

The Bailey Company is a full-service operation, providing parts and maintenance of the equipment it handles. Bailey representatives also provide, at no charge, a consultancy to clients who want to assess their needs for material-handling equipment. The firm's technical support includes training programs for lift-truck operators, as well as training sessions for clients' on-site maintenance personnel.

Bailey Company service trucks are no more than one hour away from any client. The firm employs a team of more than 130 service technicians. It operates 95 quick-response service vans equipped for on-site repairs, as well as a fleet of 14 heavy-duty machine transport haulers. The Bailey Company and its customers are linked by computer to the enormous Caterpillar parts inventory. Typically, parts for emergency repairs can be at a customer's site the morning after the parts are requested.

Years of experience combine to spell successful operations for Bailey Company management. From left are Gordon Morrow, president; Bert Bailey, executive vice-president; Fred Osborne, vice-president/sales; and Charlie Bebout, vice-president/product support.

Reemay, Inc.

Few companies practice diversification to the degree that Reemay, Inc., does. To a large extent, it is the company's manufacturing process—allowing variations of two products—that accounts for this diversification.

Reemay manufactures two nonwoven spun-

industrial products include Typar® spunbonded polypropylene, Reemay® spunbonded polyester, Tekton® spunbonded polypropylene, Typelle® needle-punched fabrics, Typar® Housewrap, Typar® Landscape Fabric, Typar® Landscape Fabric for Professionals, Typar®

In settings from the most pastoral to the most urban, products made of Reemay and Typar maintain the delicate balance between modern man and the environment.

bonded products—Typar® and Reemay®—that are modified into 900 various forms. These, in turn, are sold in 20 different markets through more than 650 direct customers in 35 countries worldwide, making Reemay truly a global manufacturer influencing millions of lives daily.

As one of the leading nonwoven firms in the world, Reemay, Inc., transforms its array of fabrics with imagination and research into a variety of end products, each designed to have unique characteristics and to perform different functions. These product transformations have proven successful time and again for Reemay's customers.

A list of Reemay's branded consumer and

Geotextiles, Reemay® Tobacco SeedBed Covers, Biobarrier® Root Control System, and Reemay® and Typar® filtration media. Reemay's products are used in a variety of other applications, such as automotive components, primary carpet backing, furniture and bedding construction, apparel interlinings, and agricultural crop covers, as well as in other industries such as aerospace, telecommunications, and medical research.

All Reemay, Inc., products are manufactured using carefully monitored spunbonding processes. Continuous filaments of polyester (in the case of Reemay products) or polypropylene (in the case of Typar products) are thermally bonded into fabric sheets that can be

Ultrafine membrane filters of Reemay enable desalination plants—from the Colorado River to the Middle East to offshore ships and submarines—to work the modern miracle of making salt water pure enough to drink.

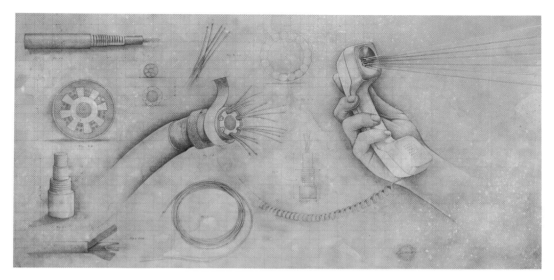

LEFT: With the addition of a superabsorbent coating, Reemay protects delicate fiber-optic cables from damage during manufacturing and from moisture after installation.

BELOW: Aircraft manufacturers use Reemay as a carrier for room-temperature-curing adhesives that "weld" aluminum components together for lightweight rigidity.

needlepunched or otherwise further processed. Reemay's flexible manufacturing process permits "adding value" to its fabric through different techniques: color sealing, printing, and chemical impregnation.

Reemay is the sixth-largest manufacturer of nonwoven fabrics in the United States (seventh in the world) and a member of the International Nonwovens and Disposables Association (INDA). More than 500 people are employed at its plant located in Old Hickory, Tennessee, north of Nashville.

Reemay, Inc., is constantly discovering new and unique uses for Reemay® and Typar®. Sometimes these new developments are a result of joint efforts by Reemay and its customers. However, the formula for industry leadership is always the same: technical excellence, creativity, and the entrepreneurial drive to quickly respond to the needs of the customer.

Reemay's innovativeness and flexibility yield products that entire industries rely upon daily. Today Reemay's problem-solving contri-

butions range from the seemingly ordinary, such as carpet backings and garment linings, to the exotic, such as support membranes in genetic research and aerospace structures, as well as sophisticated root-control systems.

LEFT: Packaging made of Reemay or Typar can be rugged but lightweight, flexible enough to conform to its contents, waterproof or water-permeable, and sturdy enough to protect structural steel.

DET Distributing Co.

A distributor of Miller Brewing Co. products, its flagship brands, as well as products of other nationally and regionally known beverage companies, DET Distributing Co. is a family-owned business with a reputation for excellent service, community involvement, and a responsible business philosophy.

It is perhaps the latter two attributes—community involvement and a responsible business philosophy—that set DET Distributing apart. Under the leadership of owner Fred Dettwiller, DET Distributing has played a

G. Frederick Dettwiller II (right), president of DET Distributing Co., and his son, George F. Dettwiller III.

major role in sponsoring community festivals that celebrate the Nashville spirit and raise thousands of dollars for Nashville charitable and cultural organizations. The Nashville Area Junior Chamber of Commerce has awarded DET Distributing a special citation for its extraordinary involvement in community affairs.

And this commitment to the community extends beyond company involvement to the personal involvement of Fred Dettwiller. He serves as chairman or board member of many citywide and statewide charitable and civic organizations, including the Nashville Convention Center Commission and The Tennessee Human Rights Commission.

Fred Dettwiller is also a firm believer in the responsible use of the products his company distributes, and it regularly conducts alcohol education seminars, distributes literature, and underwrites other programs in this area.

DET Distributing was founded in 1951 by E.E. Dettwiller, father of the current owner, as a distributor of Schlitz and Sterling beers. The elder Dettwiller, now deceased,

The DET Distributing Co. management team (left to right): Jim Gilliland, director of operations; David Earls, director of sales; Fred Dettwiller, president; and Keith Stanton, director of finance.

got his start in the beer business in 1933—the year Prohibition was repealed—when he began as a route salesman with the Tennessee Brewing Co. in Memphis. Years later, when he left to start his own company in Nashville, he had become vice-president and general manager of the Memphis firm.

Unlike the traditional family business, where the son or daughter works in the business until they take over management from the parent, Fred Dettwiller took a different, more ambitious approach. In 1956, at the age of 23, he started his own beer distributorship in Clarksville, with his cousin Bill Carter, called Cardett Distributing Co. Three years later he bought out his cousin to run the business himself. In 1973, upon the death of his father, Fred Dettwiller bought the family business from his father's estate and ran both businesses until 1982, when he sold Cardett and moved back to Nashville.

When Fred Dettwiller purchased DET Distributing in 1973, it employed 23 people and operated from cramped quarters on Clinton Street. Today it has 115 employees and operates from a large, climate-controlled building in MetroCenter. The firm currently distributes beverage products in nine Middle Tennessee counties and creates a local economic impact at the retail level of $54 million. Under Fred Dettwiller's leadership, the company has nine times been named a Miller Master—the highest honor from Miller Brewing Co.—and is regularly listed as one of Nashville's top 100 private companies.

As DET Distributing Co. continues its strong tradition of business success and community involvement, Fred Dettwiller is grooming his son, George Frederick Dettwiller III, to take over the reins of the family business.

Photo by Bob Schatz

Business And Professions

Greater Nashville's business and professional community is known throughout the South for its high level of ability and expertise.

Photo by Rudy Sanders

Kraft Bros., Esstman, Patton & Harrell

Widely acknowledged today as one of Nashville's premier accounting firms, Kraft Bros., Esstman, Patton & Harrell began modestly as a one-man office. In the more than three decades since the firm first opened its doors in 1958, it has expanded to include a professional staff of more than 100 people, including 11 partners.

Since day one the firm's dedication to providing high-quality personal service to each of its clients has remained constant. Because of this dedication to and delivery of highly personalized service, Kraft Bros. can point to an achievement that is quite unusual in the accounting business: ongoing relationships with a number of clients that have lasted for more than 30 years. Truly, Kraft Bros.' growth and success has been intertwined with the modern development of Nashville and Middle Tennessee.

From its inception the firm has been located in the heart of downtown Nashville. In November 1988 Kraft Bros. moved into more than 20,000 square feet of offices on the 11th and 12th floors of Parkway Towers at 404 James Robertson Parkway, a building with a panoramic view of the downtown business district and the seat of state government.

Kraft Bros. is justifiably proud of its history of participating not only in the economic growth of Nashville but also of its work toward the betterment of the community at large. Two partners in the firm recently have received the Public Service Award from the Tennessee Society of Certified Public Accountants.

In 1982 Kraft Bros. merged with Bone, Hundley & Sanderson of Columbia, Tennessee. At that time Bone Hundley was the oldest practicing firm in the Middle Tennessee area. With this acquisition Kraft Bros. became the largest locally owned and continuously practicing accounting firm in Middle Tennessee.

Most of the firm's clients are closely held

businesses with sales volume ranging from fewer than one million dollars to more than $100 million. While the company has grown substantially and is large enough to employ litigation support, tax, computer, human resource, and other specialists, it is still small enough to maintain the kind of close, personal attention that has resulted in so many long-term relationships with clients.

As a full-service CPA firm, Kraft Bros. has evolved beyond general audit, tax, and accounting services. The company has special expertise in serving the medical profession, the metals industry, the construction business, and warehousing and distribution, as well as government, real estate, and banking. As an early player in the Nashville music industry in the late 1950s, Kraft Bros. took the opportunity to work in specialty areas such as recording contract negotiations and audits of publishing royalties.

One of the firm's fastest-growing divisions is Human Resource Development Services, which includes personnel management, organization development and training, compensation and benefits, and executive search and staffing. Kraft Bros. has helped clients to attract, hire, and retain key executives; to structure compensation and benefit packages; to set up training programs; and to comply with the modern maze of personnel policies, many of which are now legally mandated. Another new and rapidly growing specialty is support for litigation. And, as another service, Kraft Bros. will evaluate its clients' computer systems and suggest methods for more efficient management using today's best technology.

Kraft Bros. is also involved in personal financial planning, including income tax analysis, estate planning, sheltering income, investing for growth, and planning for college funds or retirement. The firm tailors a financial plan for each individual and then carefully monitors implementation of that plan. Kraft Bros. is the acknowledged leader in Middle Tennessee in fee-only personal financial planning.

In 1984 Kraft Bros. founded LINC (Licensed Individual Network of CPAs), an organization that seeks to bring individual accountants together to refine their skills in personal financial planning. The organization conducts seminars nationwide and publishes *Fiscal Fitness*, a newsletter that focuses on the newest information on laws and trends that affect the specialty of personal financial

planning.

The firm is a member of BKR International, an international association of established and respected independent CPA firms and chartered accountants. As a member of this group, the firm maintains its autonomy but can also provide reliable expertise internationally for clients whose operations and interests extend beyond Middle Tennessee. Membership in BKR also provides access to resources only available to international accounting firms. Member firms are selected by their peers in BKR on the basis of reputation, clientele, performance, and professionalism. Each member firm is reviewed periodically to ensure that its practice is being conducted according to BKR's standards.

The professionals at Kraft Bros. are problem identifiers and problem solvers. The firm approaches each client on a proactive basis so the firm can act not only when clients call but before they call, to avoid problems. The firm is Nashville owned and operated, so clients can have confidence that decisions regarding their businesses are made by professionals they know and trust. Kraft Bros., Esstman, Patton & Harrell is proud of its history of meeting clients' expectations through its policy of carefully and patiently listening to and understanding client needs.

ABOVE: *Kraft Bros. offers an extensive computer service operation, including consulting, training, and support systems to solve the most complex business needs. KB computer services exhibit a proactive attitude, going beyond the resolution of existing problems.*

LEFT: *Kraft Bros. has long been an innovator in developing new specialties to meet an expanding market. Two wholly owned subsidiaries—LINC and Executive Development, Inc.—provide personal financial planning and human resource consulting, and are examples of the firm's commitment to being a full-service facility.*

Barcus Nugent Consulting

Barcus Nugent Consulting was formed in 1985, when computer systems consultants Sam Barcus and Terry Nugent set up shop in a spare room of a colleague's office. For some months the two men worked as independent consultants for Middle Tennessee businesses, using their computer, accounting, and business skills to improve operations within their clients' companies.

Within a very short time Barcus and Nugent had clients ranging from businesses that had been in Nashville for generations to new en-

their extensive technical expertise with their knowledge of business operations, the principals at Barcus Nugent have assisted their clients in gaining such advantages. Barcus Nugent Consulting has the resources to stay up to date and the knowledge to evaluate and install information systems, as well as the ability to train clients' personnel to properly use these new tools.

Barcus Nugent Consulting has found that few businesses are completely satisfied with their current computer systems. The system

Barcus Nugent consultants work in their design room to plan projects as a team.

trepreneurial enterprises. A rapid increase in demand for the company's services made necessary a move into much larger offices in Nashville's Parkway Towers. Since the move the firm has continued its rapid—but controlled—growth at a rate of 30 to 40 percent per year.

Computers have become an essential competitive tool in the business world. Many companies have gained clear competitive advantages through the efficient use of advanced information systems. By combining

may not perform as advertised, staff may not be adequately trained, or needs might have changed since the system was installed. New computer products come on the market daily, and applications are mushrooming. Barcus Nugent's years of experience in management consulting, computer system implementation, accounting, finance, and operations make the firm able to solve its clients' information system problems so the clients can operate more efficiently—and competitively.

Whether a company is considering the

purchase of its first computer system or wants to enhance the performance of its existing system, Barcus Nugent Consulting offers a range of services that will fit that company's needs. The firm's services include hardware and software selection, feasibility studies for in-house equipment, office automation analysis, system design and implementation, long-range data-processing planning, data security and access reviews, computer-based financial modeling, and microcomputer training. The strong accounting and business background of Barcus Nugent's staff enables the firm to take on the design, implementation, and evaluation of systems in financial and operational areas.

Barcus Nugent Consulting takes pride in what Sam Barcus calls a "craftsman's approach" to its work. Just as a master craftsman must learn the subtle nuances of his medium, whether it is the grain of wood or the shine of a tile's glaze, the consultants at Barcus Nugent must learn the nuances of blending modern technology with the unique needs of each client, says Barcus.

The firm offers three service groups to help clients effectively choose, implement, and use information technology: systems planning, design and development, and project management.

The firm's systems planning service group helps clients identify problems and find ways to correct them. Barcus Nugent consultants diagnose and analyze current system problems or define potential system needs and requirements. The analysis results in a system "blueprint" and a checklist for building effective information systems.

The design and development service group designs and builds information systems that meet clients' specific information needs. This can be an independent activity or the next step after system diagnostics. Firm members design system specifications, select hardware, and program and implement software, resulting in a new or modified information system.

The project management service group provides supplemental on-site assistance as required by a client. When specific financial or operational expertise is needed, a member of Barcus Nugent's staff may become a member of the client's management team on a short-term basis. Or, during a system installation or makeover, a Barcus Nugent staffer may serve temporarily as a client's project manager.

In years to come Barcus Nugent Consulting intends to strengthen its position in a rapidly expanding field and build on its fine reputation by seeking the best, most experienced consultants and carefully matching them with clients to best serve the clients' needs.

Principals in Barcus Nugent Consulting (from left): Sam Barcus, Nancy Britt, and Terry Nugent.

William M. Mercer, Incorporated

During the past half-century businesses have looked increasingly to independent advisers to help with the complicated field of employee benefits and compensation. At the forefront in this field is William M. Mercer, Incorporated, the world's leading actuarial, employee benefits, and compensation consulting firm.

The ability to expertly research and analyze technical issues is crucial for benefits consultants today.

Mercer was formed through two mergers. In 1984 William M. Mercer Incorporated merged with Meidinger Inc. Three years later A.S. Hansen, Inc., joined the organization. Today the firm has more than 3,000 employees in 50 offices in the United States, and the Mercer companies worldwide make up the world's largest consulting firm in the field, with more than 6,000 employees and offices in 20 countries. The firm serves more than 16,000 clients in business, industry, government, and nonprofit operations.

The Tennessee operation includes consulting offices in Nashville and Memphis. The Nashville office is made up of actuaries, health and welfare consultants, compensation specialists, defined contribution specialists, attorneys, and various support personnel. The office serves a wide variety of public and private employers in the Middle Tennessee area.

Mercer is part of Marsh & McLennan Companies, Inc. The Marsh & McLennan family includes specialized financial consulting and service firms in a number of related fields. One of its subsidiaries, Marsh & McLennan, Inc., is the world's leading insurance broker.

Mercer's services can be grouped in five areas: employee benefits, design, and administrative services that attract, motivate, and retain employees; health care cost management that helps employers ensure that employees receive quality health care at a reasonable cost; compensation systems, organizational design, and job planning, which help employers stay competitive in the marketplace by rewarding employees for performance; communication and human resource management that help employees identify with management objectives; and funding and financing of compensation and benefit programs that offer employers the widest range of options and help control costs.

As Middle Tennessee's employee population continues to grow, William M. Mercer, Incorporated, expects to continue its growth. The company has consistently allocated resources to the research and development of new systems and tools that will extend its range of consulting services. These are plans made for the long term, as the company continues to build on its solid foundation in Nashville for the successful years ahead.

Mercer knows that understanding an organization's objectives is critical to providing sound consulting advice.

Corroon & Black Corporation

Corroon & Black Corporation, with its worldwide network of risk-management and loss-control specialists, is an acknowledged international leader in the business of insurance services. The corporation's Nashville operation employs a work force that constitutes the largest concentration of Corroon & Black resources worldwide.

Beyond traditional risk transfer, Corroon & Black professionals provide a multitude of services, including self-insurance consulting and third-party administration. The corporation has a distinguished track record in developing products that respond to the sophisticated risk-management requirements of specific business sectors, including the construction

& Black's formal name introduction in Nashville occurred in 1976, when the firm merged with Synercon Corporation, a leading regional insurance holding company. That merger was recognized as the largest in the history of the U.S. insurance-brokerage industry.

Today, under the experienced leadership of Richard M. Miller, chairman and chief executive officer, Corroon & Black Corporation enjoys an unparalleled reputation in the insurance-services industry.

Corroon & Black's commitment to Nashville and Middle Tennessee is stronger than ever. A new corporate complex will be opened in the city in the early 1990s. This is

Corroon & Black Plaza Phase I, the firm's new Nashville corporate office, is scheduled to open in October 1991.

and marine industries. Corroon & Black also specializes in benefits consulting, with a particular expertise in actuarial and administrative services for pension and health plans. The company also provides services in the areas of employee compensation programs, employee communications, and financial and estate planning.

Corroon & Black's Nashville roots can be traced back more than 100 years to a predecessor firm that served the insurance needs of businesses transporting goods along the Tennessee and Cumberland rivers. Corroon

expected to further enhance the sharing of resources by bringing together the many Corroon & Black professionals who work in the areas of property, liability, marine, fidelity, surety, and individual and group benefits.

With its wide array of products and services, managed by a worldwide network of experienced professionals, the firm looks confidently to the future. At the dawn of a new decade, Corroon & Black Corporation and its Nashville operations continue to build on a proven ability to respond to changing needs in a changing world.

American General Life and Accident Insurance Company

For years Nashville's skyline has been framed by two landmark skyscrapers—the National Life Center and the Life and Casualty Tower—twin pillars that were the longtime homes of competing local insurance giants.

These buildings still stand tall, but the two companies now occupy the American General Center, the 31-story building that once was the headquarters of National Life, and both have been acquired by American General Corporation, the nation's fourth-largest stockholder-owned insurance company. The two once locally owned giants are quickly becoming one under American General Life and Accident Insurance Company—a name that is becoming as familiar to Nashvillians as National Life and Life and Casualty.

Through a series of consolidations, American General Life and Accident has become the headquarters for other American General insurance operations, including Equitable Life of Virginia, American General of Oklahoma, and American General of Delaware.

Despite such growth, people around the company still recall past leaders, men such as founding chairman C.A. Craig of National Life and A.M. Burton of Life and Casualty, as well as their successors. The names, faces, and accomplishments of National Life and Life and Casualty are a rich part of Nashville's history, and those images will remain prominent in the memories of Nashvillians for years to come.

In 1922 executives of National Life joined the fledgling radio industry by launching a station that would be a vehicle of goodwill for the company. The call letters, WSM, were chosen to reflect the stability and prosperity of the insurance company: the letters stood for "We Shield Millions." WSM later introduced a barn dance program that evolved into the Grand Ole Opry.

For many years National Life meant the Grand Ole Opry as much as it did insurance. National Life and Life and Casualty also meant WSM television and radio, Opryland USA, and WLAC television and radio.

Many Nashvillians recall Life and Casualty's investment in Nashville's growth through its building in 1957 of the city's first skyscraper, the L&C Tower. The 25-foot-high "L&C" letters atop the building were used for decades to report weather conditions for the Nashville area by the changing of their colors.

Over the years the two companies were recognized for their community involvement at all levels. American General Life and Accident continues the tradition of community involvement started by its corporate predecessors. In 1988 the firm received Nashville's Corporate Philanthropy Award in recognition of volunteer work done by its employees.

Throughout the year the American General Center itself serves as a giant bulletin board for the city, spelling out messages that are visible for miles around downtown Nashville. During

The American General Center is American General Life and Accident's Nashville corporate home.

the city's now nationally acclaimed Summer Lights festival—a four-day downtown fete that features virtually all styles of music, as well as the visual and performing arts—the lights in the building spell out "Arts, Music, Dance" in letters several stories high. During the winter holiday season the building shines with the words "Peace On Earth." Fund drives for local charities are frequently announced on the building, reminding everyone who sees the Nashville night skyline of the needs of the community and of American General Life and Accident's commitment to its community.

Today, as a subsidiary of the American General Corporation and headquartered in Houston, American General Life and Accident is part of one of the nation's largest insurance and financial services organizations, with assets of more than $30 billion and shareholders' equity of more than $4 billion. American General Life and Accident provides individual life and health insurance and personal insurance services to more than 12 million policyholders.

The firm is one of Nashville's largest employers, with a home office staff of approximately 1,500 people providing support for more than 6,000 full-time sales and sales management representatives in 250 district offices in 25 states. More than one-quarter of the home office employees can boast having 25 or more years of service with the company.

Shareholders' equity, earnings, capitalization, stock prices, and other measures of financial growth have been notable since American General's acquisition of National Life and Life and Casualty.

Even so, there is more to American General Life and Accident Insurance Company than great size and financial strength. It is an organization with proven performance in serving the needs of individuals, families, and businesses on a personal basis—proven performance that its policyholders and the city of Nashville can depend on in the future.

Nashville's first skyscraper, the L&C Tower, is a rich part of American General's—and Nashville's—history.

During Nashville's annual Summer Lights festival, the American General Center serves as a 31-story bulletin board for the city. Photo by Gary Layda

Dearborn & Ewing

Dearborn & Ewing was founded in 1972 through the merger of two small but well-established Nashville law firms. Since that time the firm's practice has expanded to become a full-service civil practice that can meet the needs of any client.

The firm was established with one overriding goal in mind that continues to guide its practice today: The lawyers at Dearborn & Ewing are committed to deliver the highest-caliber legal services in the most timely and efficient manner.

Dearborn & Ewing is organized into law-practice sections: banking and commercial law, litigation, real estate and environmental law, business, and trusts and estates. The firm has also developed a number of less formal practice areas in which lawyers combine their talents to address clients' special needs.

The lawyers in Dearborn & Ewing's banking and financial institutions section have the ability and experience to manage any type of lending transaction that arises in the Mid-South. The firm is particularly experienced in real estate finance, from acquisition and development loans through construction lending and permanent financing for properties. Dearborn & Ewing has served as primary counsel to the leading banking and financial institutions in the Mid-South.

Real estate law has traditionally been a significant portion of Dearborn & Ewing's practice. Attorneys in this practice area have extensive experience in advising individuals, financial institutions, and publicly and privately held entities in all areas of real estate operations. The members of this practice area of the firm have represented some of the largest real estate acquisition, management, syndication, and development firms in the United States.

Environmental law is a relatively new and rapidly evolving area of the law to which Dearborn & Ewing has committed substantial resources. The firm counsels lenders, developers, sellers, and purchasers as they address environmental problems that affect a variety of transactions, ranging from the acquisition of real estate or a business to foreclosure of a loan.

Manufacturers and other businessmen call on the lawyers in this area to resolve issues involving the use, production, transportation, storage, and disposal of hazardous materials; the cleanup of hazardous waste sites; and the discharge of regulated substances into the water or the air. The firm also assists clients in obtaining necessary permits and monitors compliance under federal and state environmental laws. To help its clients stay current on issues in the environmental law area, the firm publishes *Environmental Law Update*, a quarterly newsletter highlighting issues and developments in environmental law.

Dearborn & Ewing regularly represents and counsels clients in the formation and operation of business organizations, including corporations, general and limited partnerships, and joint ventures. Attorneys analyze and provide advice concerning business plans, financial structures, relationships among investors, and short and long-term compensation for key personnel. The firm's clients include publicly and privately held corporations, partnerships, and individuals, including management groups and outside

Experienced leadership generates creative solutions to complex legal issues.

LEFT: An understanding of technology adds to the development of environmental practice skills.

FAR LEFT: The firm's entertainment practice includes the protection of the business interests and intellectual property rights of composers, recording artists, and publishers.

investors participating in the leveraged buy-outs of existing businesses. The firm also publishes a quarterly newsletter, *Business Law Update*, highlighting issues and developments in business law.

Attorneys in the firm's trusts and estates section devote a substantial amount of their time to estate planning, giving advice with respect to drafting of wills, trust agreements, and probate and estate administration problems.

Through the merger of Gillmor, Anderson & Gillmor with Dearborn & Ewing, the firm now has attorneys with a national reputation specializing in health care law. These attorneys, working with the firm's specialists in the area of business and finance, will continue to provide representation to clients in all aspects of health care business, as well as financial institutions involved in the extension of credit to these types of entities.

The firm is widely recognized for its expertise in both traditional municipal finance and private activity bonds. Dearborn & Ewing's reputation in the public finance field has grown due to its involvement in successful development of innovative financing vehicles to meet the needs of public clients in fluctuating capital markets. Members of the firm have been involved in innovative methods of finance that later became widely used, including the nation's first tax-exempt variable-rate general obligation bond.

With its broad experience in international transactions, Dearborn & Ewing occupies a unique position among Nashville law firms. For most of its history, the firm has been active in both export and import transactions, representing U.S. businesses and foreign companies.

Dearborn & Ewing represents a number of clients in the areas of entertainment and intellectual property law. Attorneys give advice to persons and companies in the entertainment industry and related fields, including recording and performing artists, composers, publishers, and producers. This work includes negotiation

and preparation of contracts related to the creation, exploitation, purchase, and sale of musical compositions, sound recordings, and other related property rights. The firm has built on its experience in the entertainment industry to serve the needs of clients who produce other types of intellectual property, such as computer software.

Dearborn & Ewing's litigation section is experienced in every area of civil law. The firm represents a wide variety of clients in all areas of litigation in both federal and state courts, as well as representing clients before administrative and arbitral tribunals. As in other areas of practice, the litigation area has evolved with the changing needs of clients. The banking and savings and loan crises of recent years involved this section in directors' and officers' liability litigation. Although primarily defense oriented, the firm's litigation practice has expanded to include plaintiffs' cases in tort and business disputes.

A friendly, yet professional atmosphere generates a progressive attorney-client relationship.

Building
Greater Nashville

Designing, developing, and managing property, real estate and building firms shape the Nashville of tomorrow.

Photo by Rudy Sanders

Maryland Farms

M aryland Farms was once home to a world-famous American Saddlebred horse farm nestled in the lush rolling hills of Middle Tennessee. Today Maryland Farms is a premier office park in the progressive suburban community of Brentwood, Tennessee.

Good location and strong, experienced leadership have made Maryland Farms the unique leader among Nashville's office parks. Located south of Nashville, Brentwood affords easy access to three major interstates (65, 40, and 24) and the new Nashville International Airport. Maryland Farms Office Park is set in the heart of the executive housing area that extends south from Nashville through Brentwood.

Progress in Brentwood has been encouraged by its central location between Nashville, Franklin, and the new General Motors Saturn operation in Spring Hill. Excellent schools, public parks, libraries, convenient shopping, and numerous services add to the quality of life-style in this family-oriented community.

From its origination in 1974 as an office park development, Maryland Farms has shared in the steady growth of the Middle Tennessee economy. Many of its original tenants have also shared in that growth and have found Maryland Farms a great place to experience expansion.

The pace of economic growth in Middle Tennessee increased in the late 1980s, but Maryland Farms is only 40 percent completed, with choice real estate and leasable office space still available for corporate users.

A stable and experienced management team at Maryland Farms is a major strength of the development. Five partners whose individual accomplishments are among the most distinguished in American business oversee the development of Maryland Farms. Four of these partners have been involved in the development since its beginning, and one is a member of the Ward Family, which owned the original Maryland Farms.

Businesses located in Maryland Farms reflect the vigorous, diverse Middle Tennessee economy. Small independent companies with a local or regional focus flourish beside large national and international corporations. To the workers who come to Maryland Farms daily, it is much more than just a growing 1.5-million-square-foot office park. Maryland Farms offers an active, vibrant environment to complement today's executive life-style. As home to local government offices, a library, retail shops,

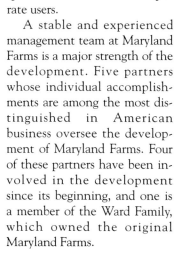

The appreciation of natural beauty and its preservation makes Maryland Farms the unique leader among Nashville's office parks.

Maryland Farms Racquet and Country Club offers 7 indoor and 16 outdoor tennis courts, 5 racquetball courts, indoor and outdoor swimming pools, exercise equipment, excellent catering, plus dining and meeting facilities for conferences and conventions.

The natural beauty of Middle Tennessee provides an appropriate setting for an office park where unobtrusive buildings of brick and glass blend with ample green groves of majestic trees and well-groomed beds of flowers.

Despite all that has changed, Maryland Farms still stands as a testimony to the time-honored values that made it a successful family-owned horse farm for more than 30 years. Management of Maryland Farms insists that every effort is made to preserve and enhance the natural beauty of the area. Workers and visitors appreciate the commitment Maryland Farms makes to creating a fresh, appealing atmosphere. Nestled in 400 acres of beautiful Tennessee countryside, Maryland Farms maintains an environment that is peaceful yet progressive.

medical facilities, child care services, financial institutions, dining opportunities, and a state-of-the-art athletic center, Maryland Farms is a community within a community.

At the center of the office park, the

Walter Knestrick Contractor, Inc.

Founded in 1969, Walter Knestrick Contractor, Inc., is one of the top ranked general contractors in Nashville and the Southeast. The company is uniquely experienced in commercial, industrial, and retail construction, including interior construction and historic restorations. Having earned a reputation for excellence, the firm is known for its attention to quality and detail.

The Knestrick projects list reads like a roll call of southeastern commerce. Many well-known national companies, including American Airlines, Reynolds Aluminum, Du Pont, Service Merchandise, Robert Orr-SYSCO, and Bendix, join with homegrown operations such as First American National Bank, Nashville Gas Company, Peterson Tool, Surgical Care Affiliates, the University School of Nashville, and Ingram Industries to make a

The Walter Knestrick commitment to quality construction can be seen at the Bendix-Jidosha Kiki Corporation headquarters and manufacturing plant (BELOW) and at the Nashville Gas Company corporate headquarters (ABOVE).

list of more than 1,000 completed projects.

Walter Knestrick Contractor has built several important facilities for many of the Japan-based companies that have established operations in Middle Tennessee. The firm has completed jobs at the Bendix-Jidosha Kiki manufacturing plant, Bridgestone's tire-manufacturing plant, and Nissan's automobile

factory in Smyrna.

The company takes pride in the many ongoing relationships it has developed with clients. This has resulted in a large repeat business, such as the construction of Service Merchandise retail stores, YMCA facilities, Keystone Foods plants, Surgical Care Affiliates surgery centers, and Tractor Supply Corporation's headquarters and many of its stores.

Knestrick has successfully met the challenge of many customers for differing requirements within the same facility. In a construction job for Nashville-based Ingram Industries' corporate headquarters, Knestrick's work included standard offices, an employee cafeteria, a computer center, executive dining rooms, and executive suites.

Walter Knestrick Contractor also has garnered acclaim for its construction in historic restorations, including the restoration of former President Andrew Jackson's home, the Hermitage. In this project, for example, the upgrading of mechanical systems in the 154-year-old home (to control temperature and humidity for preservation of original furnishings, wallpaper, and books) had to be completed in a way that would maintain the historic architectural details.

Achieving the exacting demands of customers is Knestrick's primary purpose. In the case of Nashville Gas Company's corporate headquarters, Knestrick constructed the building that later won the 1989 Modernization Excellence Award from *Buildings* magazine.

Walter G. Knestrick, the company's founder, has the philosophy that his firm must provide a good product, on time, at a reasonable price, and leave a client satisfied. He believes that if his company always meets those four goals, Walter Knestrick Contractor, Inc., will always be successful.

Lee Company

From a very humble beginning in 1944, operating out of the basement of a home and using a 1940 Chevrolet sedan as a service truck, the Lee Company has grown to be one of the top 100 mechanical contractors in the United States.

Lee Company specializes in heating, ventilation, air conditioning, plumbing, and process-piping installations. Employing more than 200 people, the firm is licensed and has performed work in 26 states. Lee Company is unique in the mechanical contracting business in that more than 70 percent of its business takes the form of design/build contracts in which its staff of four licensed mechanical engineers designs and reviews all projects and operations. Lee Company is among the top 100 privately held corporations in Nashville, and is the largest air-conditioning

and plumbing service contractor in the Nashville area. The firm is employee owned through an employee stock-ownership program.

Lee Company has one of the largest service and maintenance operations in Tennessee, with trained technicians responding to customer calls with a fleet of more than 20 service trucks. The firm services all types of heating, air-conditioning, refrigeration, and plumbing installations for commercial, industrial, and residential customers.

The service department exemplifies Lee Company's dedication to achieving complete

customer satisfaction. The firm's goal is to not only satisfy its customers before and during the job but also to service and maintain the equipment and installation indefinitely. Lee Company's growth is a result of this dedication to ongoing service; more than 80 percent of the firm's business comes from repeat customers and referrals from previous customers.

In 1983 Lee Company received national recognition by being named Commercial and Industrial Contractor of the Year by *Contracting Business* magazine. Wallace Lee, president of Lee Company, served two terms as national president of Air Conditioning Contractors of America in 1984 and 1985. The firm has successfully completed mechanical installations for companies such as Nissan, Bridgestone, Whirlpool, Fafnir Bearing, Koppers, Essex Industries, Heil Quaker, Radisson, Marriott, and Pickett Suite. Lee Company has also worked with many international clients, such as Calsonic Yorosu, Kanzaki Paper, Teksid, Kobelco Metal Products, and Asahi Motor Wheel.

Principal officers of Lee Company include Wallace Lee, president; Ted Lee, secretary/treasurer; William Lee, vice-president; and Michael Bishop, vice-president. Through its family of dedicated professionals, Lee Company will accept the challenge of excellence by meeting the needs of its customers competently, honestly, and courteously, establishing standards by which others are measured.

LEFT: Throughout the city of Nashville, Lee Company service vans are familiar wherever quality mechanical contracting is needed.

Edwards + Hotchkiss Architects, Inc.

RIGHT & FAR RIGHT: Located in Maryland Farms Office Park, a 50-year-old historic horse barn is home to the corporate office of Edwards + Hotchkiss Architects, Inc.

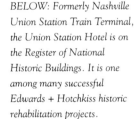

BELOW: Formerly Nashville Union Station Train Terminal, the Union Station Hotel is on the Register of National Historic Buildings. It is one among many successful Edwards + Hotchkiss historic rehabilitation projects.

I n 1975 James Edwards faced a career crossroads. He was a new partner in a two-man architectural firm, and the older partner was moving to greener pastures in California. As Edwards contemplated future career options, he consulted with his close friend Dick Hotchkiss, who was then serving as the director of design for the Tennessee Department of Conservation.

Soon afterward Edwards + Hotchkiss Architects, Inc., was built around the beliefs and personalities of James and Dick: openness, hard work, and clear purpose. Their purpose was to develop a successful architectural business that would provide creative and technically solid design services on every project, every time, for every client.

Today the company is a larger model of the original. Edwards and Hotchkiss have proven that the way to secure the future of their firm is to create a work place that supports their original concept of an open and creative environment.

Edwards and Hotchkiss are not simply the principals, but part of a company team effort. Everyone at the firm shares in the responsibilities and duties of the office, and this team intends to endure by maintaining its high regard for clients and co-workers. Edwards + Hotchkiss manages a hefty work load with large-firm capabilities and small-firm flexibility and personality. All the members of the organization have the experience and know-how to manage their own companies, but prefer being a vital part of the Edwards + Hotchkiss team.

The consistent growth in the area—particularly the Nashville relocation of many national and international corporations—has played an important role in the success of Edwards + Hotchkiss. After 15 years the firm has a proven track record, with previous clients providing almost 60 percent of the company's annual work. Edwards + Hotchkiss has extensive experience in lodging, restaurant, historic rehabilitation, retail,

office, recreation, multifamily projects, and single-family residential design. It also has many religious, educational, institutional, and manufacturing/processing projects to its credit.

Much of Edwards + Hotchkiss' work has required unique design services for specialized projects, such as historic rehabilitation of a 100-year-old train station into a hotel, facade renovation and structural remedy in a 28-story office tower, conversion of a partially completed condominium development into a luxury resort, and design of a high-rise office and housing facility around an elevated steel water-storage tower.

Other projects include placement of a camp housing and meeting facility among several old and architecturally significant buildings, revising a prototypical restaurant design to satisfy the neighborhood historic district review board and the owner, designing a major image-enhancement scheme for more than 300 similar buildings, and converting a 50-year-old, wood-frame horse barn into an office facility.

Currently, Edwards + Hotchkiss' projects are represented from Washington to Grand Cayman and from Boston to Southern California. The firm has professional architectural licenses and projects in 35 states.

The firm's awards and recognitions include the Golden Gate Award for community contribution for The Horse Barn at Maryland Farms in Brentwood; the Federal Design

Achievement Award and Presidential Design Award for Union Station Hotel in Nashville; a *Tennessee Architect* feature for use of natural energy in recreational architecture for Bethany Hills Camp and Training Center near Kingston Springs, Tennessee; a design competition award for Highland Ridge III Office Tower in Nashville; regional winner of *Metropolitan Homes'* Home of the Year Award for the Key/Whitehead residence in Nashville; and a nationwide design competition award for Mrs. Winner's restaurants.

Edwards + Hotchkiss has quietly secured its place as one of Middle Tennessee's leading designers of custom residences, currently designing more than 20 homes each year. The firm can satisfy any home owner's design desires, whether it be a Mediterranean villa, an English country manor, a California contemporary, or a mountain lodge, whether the site is valley, lakefront, or hilltop.

Although not afraid of further expansion, the firm expects that the personnel level of the past four years (around 20 employees) will continue for some time. The company, however, will endure long after the original namesakes have served their time. Perhaps Rick Stoll and Colleen Atwood will be the managing principals in another 15 years; they have already accumulated more than 20 years of tenure and 30 years of experience between them, and both serve as directors on the board. In years to come the company will likely feature the work of Mike Ireland, Beth Cashion, or Eric Powers—now project architects for the firm—or the work of any one of many younger apprentices making their way into leadership roles at Edwards + Hotchkiss Architects, Inc.

Edwards + Hotchkiss designed these beautiful private residences situated in the Oak Hill (TOP) and Brentwood (BOTTOM) communities.

LEFT: The Highland Ridge III Building, corporate headquarters for Centex-Rodgers Construction Co., is located adjacent to Nashville International Airport.

BELOW: The corporate office building for Cracker Barrel Old Country Stores, Inc., is a one-story office building on a heavily wooded rural site.

Gould Turner Group, P.C.

ABOVE: The corporate offices of the Gould Turner Group, P.C., are located in Palmer Plaza and are a testament to the firm's excellence in design.

LEFT: Palmer Plaza, designed by the Gould Turner Group, P.C., in 1986, adds perspective to the Nashville cityscape.

Whether designing health care facilities or high-rise offices, the Gould Turner Group, P.C., brings a unique blend of style and pragmatism to each of its clients.

Steve Turner and Mike Gould, the principals in the Nashville-based and nationally recognized architectural firm, have built a reputation based on their special blend of imagination and practicality. Turner sums up his company's philosophy this way:
"If a space does not function well, neither will the people who occupy it."

Since Turner and Gould moved to Nashville some 15 years ago, their firm has seen—and contributed to—the rapid growth of Nashville and its skyline. During that time the Gould Turner Group has been favorably recognized as a diverse and multifaceted design firm. Its designs include projects for major health care companies, including Nashville-based Hospital Corporation of America and HealthTrust Inc. From magnetic resonance imaging (MRI) facilities to the popular new home-like birthing rooms, the Gould Turner Group has demonstrated its ability to take even the most technologically sophisticated equipment and put it in an environment that is both user friendly and comfortable.

The company's approach is straightforward, based on thorough research at the beginning of a project. Before a single line is drawn, Turner and Gould make sure they understand the needs of the client, the limitations of the space, and all regulatory restraints. This emphasis on up-front planning assists the architects in designing a functional space, and it gives their clients an accurate cost estimate while helping to put their design needs into perspective.

The Gould Turner Group does not rely heavily on convention when developing its designs. Instead, it uses state-of-the-art Computer-Aided Design and Drafting (CADD) to produce remarkably precise production drawings.

The particular needs of each client dictate a specific approach, and the principals at the Gould Turner Group believe that it is in matching the structure's design with its function where creativity should begin. Turner and Gould believe that it does not matter how beautiful or interesting a building is if it does not fulfill the client's needs. Gould Turner Group buildings are designed for the clients, not the architects.

But that does not mean that buildings have to be boring. This is evident in the many projects designed by Gould Turner Group, P.C. Beautiful and appealing, each design also brings its own unique solutions to the task of allocating space and solving problems.

CANTERBURY CLOSE

DERBY GLEN LANE

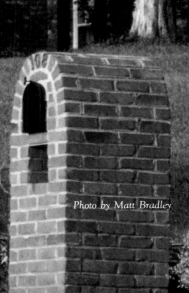

Photo by Matt Bradley

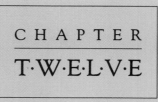

Hospitality

W hether sampling the sights and sounds of Music City U.S.A. or on business in the Wall Street of the South, travellers find convenience and comfort in the city's fine hotels.

Photo by Bob Schatz

Maxwell House Hotel

S ocial center of the Mid-South for three generations, the original Maxwell House Hotel epitomized southern hospitality. The legacy lives on in today's Maxwell House. With fine accommodations in 289 guest rooms, the newly renovated Maxwell House Hotel is building a reputation for gracious hospitality that equals that of its famous predecessor.

The Maxwell House was built by Colonel John Overton, Jr., of Travellers' Rest and was named for his wife, the former Harriet Maxwell. Its construction began in 1859 and was interrupted by the Civil War when the city was occupied by Union troops in February 1862. Opened to the public in 1869, the Maxwell House was for nearly 100 years a political, business, and social mecca, known for its cordiality and cuisine. The Christmas Day menu of 1879, for example, offered 22 meats, including Cumberland Mountain black bear, pheasant, and suckling pig. The present-day Crown Court Restaurant continues the tradition of renowned food served in an elegant atmosphere—and with a magnificent view of the Nashville skyline.

At the stately old Maxwell House at Fourth and Church in downtown Nashville, guests included such celebrities as presidents Andrew Jackson, Rutherford B. Hayes, Benjamin Harrison, Grover Cleveland, William McKinley, Theodore Roosevelt, William Howard Taft, and Woodrow Wilson. Other famous guests were Thomas Edison, Henry Ford, Buffalo Bill Cody, and Tom Thumb.

Shortly before the turn of the century, Nashville entrepreneur Joel Cheek perfected a special coffee blend that became the house blend for the Maxwell House. When he began selling the coffee to the general public, he adopted the hotel's name as the brand. A world-famous slogan was born when President Teddy Roosevelt visited the hotel in 1907 and proclaimed that the coffee was "good to the last drop." Maxwell House coffee is still proudly served at the hotel today.

The original Maxwell House was destroyed by fire on Christmas Day, 1961, but today's rebuilt Maxwell House is keeping its unique spirit alive. Guests can see it in the stately grandfather clock in the hotel lobby and in the lace balloon curtains in the Crown Court. The hotel's dedication to old-time hospitality is

evident in the personalized service, the attention to detail, and the extra effort the hotel staff makes to ensure each guest's comfort. Just like its famous predecessor, the new Maxwell House strives to make everything about the hotel something special.

The writer O. Henry once described the style of service at the original Maxwell House as "attention full of exquisite southern courtesy" and went on to say that "the food was worth traveling a thousand miles for." The Maxwell House staff believes he would repeat those comments if he could visit today's Maxwell House.

The Maxwell House is just two miles north of downtown Nashville, 15 minutes from Nashville International Airport, and within walking distance of festival shopping at Fountain Square. For guests' convenience, the hotel provides complimentary transportation to and from Fountain Square and to and from Nashville International Airport. The Maxwell House is conveniently located adjacent to Interstate 265, which connects to all the interstate highways around Nashville.

Having recently undergone a $2-million renovation, the Maxwell House provides such amenities as king- or double-size beds in each room. Each room also includes an AM/FM radio with an alarm clock, a remote-control color television, a lighted desk area, and a high-back stuffed chair. Guests also have easy access to more than 300 parking spaces, all free of charge.

For meetings, conventions, and banquets, the Maxwell House boasts 12 meeting and banquet rooms totaling more than 12,000 square feet of space. Each of these rooms is conveniently located in the same area, well away from the other traffic areas in the hotel.

Maxwell House guests also have complimentary use of the hotel's health club, which includes a steam room and sauna, an oversize whirlpool, and exercise room; two lighted tennis courts; an outdoor pool and deck; and jogging trails.

In addition to the award-winning Crown Court Restaurant, with its breathtaking view of the city, the Maxwell House also is proud of Pralines, its informal restaurant that overlooks the

hotel tennis courts, and Maxwell's Lounge, a comfortable setting with friendly atmosphere and live entertainment nightly.

As it seeks to match and exceed the original Maxwell House Hotel's tradition of excellence, the staff of the modern Maxwell House wants guests to know that during their stay they can be sure that they will receive the hotel's respected service, which will not only meet but exceed its guests' highest expectations.

In addition to fine dining, the award-winning Crown Court Restaurant offers a breathtaking view of the city.

All rooms at the Maxwell House Hotel are well-appointed, featuring either king- or double-size beds, a remote-control color television, and a lighted desk area.

Loews Vanderbilt Plaza Hotel

The Loews Vanderbilt Plaza Hotel ranks highly among the nation's best hotel values.

The Loews Vanderbilt Plaza Hotel is a contemporary hotel enriched by traditional hotel hospitality and courteous, attentive service from a cordial and competent staff. The hotel is superbly located in Nashville's prestigious West End Corridor, at the hub of the city's cultural, business, medical, and educational activities.

Vanderbilt Plaza is only nine miles from Nashville International Airport, and the hotel is only one mile from Interstate 40, which connects directly with the airport. The city's downtown business district is two miles east of Vanderbilt Plaza, and Nashville's famous theme park, Opryland U.S.A., is 14 miles away, just off I-40. Cultural attractions such as Nashville's famous Cheekwood museum and gallery and the Belle Meade Mansion are both a few minutes drive west of the hotel. Just blocks away are Vanderbilt University, Vanderbilt Medical Center, and Baptist Hospital, as well as Centennial Medical Center and Hospital Corporation of America's corporate headquarters. Area shops are acknowledged as the finest in the city.

Luxurious dining is offered at Chancellor's, the hotel's premier restaurant, or at Impressions, a cafe where diners can watch pastry chefs creating delicacies in the adjoining bake shop. Entertainment and cocktails are available in Snaffles, a charming pub, or in the Garden Lounge, a split-level retreat that is a comfortable place to meet business associates and friends.

In Chancellor's, guests can choose from a fine selection of dishes that are expertly prepared by the hotel's chefs. The classic American and continental cuisine is unsurpassed, and the service is impeccable. At Impressions, hearty breakfasts, lunches, and dinners are served in informal surroundings.

RIGHT: The chancellor's Board Room is ideal for the needs of business travelers and guests.

Each of the hotel's 342 guest rooms and suites offers privacy and comfort. All are supremely lavish and spacious. Rooms and suites feature custom furnishings, bedside control panels, executive work areas, and in-room movies. Some suites are available with fireplaces. On the Plaza Levels—two special luxury floors with concierge service—continental breakfasts, hors d'oeuvres, and cocktails are served in the privacy of the Plaza Level Lounge.

The Loews Vanderbilt Plaza Hotel is ranked sixth in the United States in the Zagat Hotel Survey's category of 40 Best Hotel Values. More than 850 hotels, resorts, and spas are included in the survey; survey participants aver-

age 46 hotel nights per year. More than 4,000 frequent travelers and travel professionals responded to the Zagat's survey, as well as members of several professional associations, including Meeting Planners International, the National Speakers Association, and the Society of Incentive Travel Executives.

In addition to being rated the sixth-best hotel value in America, Loews Vanderbilt Plaza ranked first in the category of Best Hotel Buys By City, with an overall rating of excel-

The Centennial Ballroom, with an impressive 8,400 square feet of space, can be divided into three separate rooms.

lent in all categories surveyed by travelers, including Best Rooms, Best Dining, and Best Service.

Richard Markham, general manager of Loews Vanderbilt Plaza, credits his staff for the hotel's rating. "We are honored not only to have been included in the survey but to have been rated excellent by travelers from across the country. Our service staff has done an exemplary job in putting the guest first," says Markham.

The Loews Vanderbilt Plaza Hotel provides an ideal setting for a special business or social gathering. The combination of a prestigious setting, efficient facilities, and attentive, professional service are certain to make guests' meetings memorable, productive, and successful.

The hotel's main ballroom provides 8,400 square feet of space and can be divided into three separate rooms. On the Mezzanine Level,

12 additional rooms are available to accommodate smaller gatherings. Guests can count on the professional staff of the Loews Vanderbilt Plaza Hotel to attend to every detail of their meeting. Planning is thorough, scheduling is efficient, and every aspect of a special meeting is carefully supervised. Meeting planners will even have a personal representative available to work with their group from initial planning through the close of the gathering, to keep the proceedings on target.

Acknowledged as one of Nashville's finest hotels, Loews Vanderbilt Plaza Hotel is proud of its reputation as a contemporary hotel with exemplary amenities and service and unaffected southern charm.

ABOVE: The Vanderbilt Plaza's well-appointed guest rooms greet weary travelers at the end of a long day.

Sheraton Music City Hotel

The Old Southern style grand white pillars, trellised veranda, and landscaped grounds of the Sheraton Music City Hotel's exterior are complemented by all the most modern luxuries offered within.

Set in the gently rolling hills of the Nashville countryside is a hotel that reflects Music City's special character. The Sheraton Music City Hotel is built in the classic Georgian architectural style. Seeing its grand white pillars, trellised veranda, and carefully landscaped grounds, a guest can easily conjure a vision of the Old South. Through the double brass doors of the entryway is a lobby graced by fine period furnishings and rich cherrywood paneling, with the heritage of fine Tennessee manor homes spun into every detail. McGavocks II, the hotel's high-energy lounge, is a favorite of Nashvillians and visitors alike.

This classic ambience is made all the more remarkable by the hotel's gracious staff members, who are there to greet visitors as guests of honor and guide them through a pleasurable, productive visit. Friendly and professional service add to the comforts of southern living, and they are complemented by contemporary urban luxuries such as tennis courts and pools, as well as a health club and jogging trail. Fresh and graceful, exciting and sophisticated, the Sheraton Music City Hotel is proud to welcome its guests to Nashville.

The Sheraton Music City Hotel is in the

center of the urban corporate growth in Nashville. Located three miles north of Nashville International Airport, it offers a complimentary shuttle service. It is also just seven miles from Opryland USA Theme Park.

Extending its renowned hospitality to tourists and business travelers alike, the Sheraton Music City Hotel boasts 412 spacious guest rooms, each one tastefully decorated and featuring remote-control television and video checkout. Perfect for private discussion, quiet work, or simple rest, each room is appointed with such first-class amenities as a comfortable writing desk and three

telephones—very useful for business travelers. When it is time to relax, there is a comfortable sofa in each room; all rooms also have a balcony or patio with an attractive view of the pool courtyard or Nashville's gently rolling hills. Guests can enjoy any of the hotel's 56 suites—including three presidential suites—each with a wet bar and refrigerator. The Sheraton Music City also features complete handicapped facilities.

Guests can renew their vitality in a country-club setting. The hotel's full-service health club is outfitted with the finest equipment, such as Tunturi exercise bikes, rowing machines, hydro fitness equipment, a universal Power Pak 400 machine, an exercise bicycle, and a treadmill. Other amenities include indoor and outdoor pools, spa facilities, tennis courts, and a beautifully landscaped jogging trail. Individual and corporate memberships are also available to its health club, and hotel guests may use the health club free of charge.

Meeting and banquet facilities totaling nearly 26,000 square feet are conveniently located on the lobby level of the hotel. Nashville's Cordial Manor has the perfect setting for many types of meetings and events. The spacious ballroom foyer allows easy access for exhibits. There are 15 meeting rooms and six hospitality suites. The Plantation Ballroom contains more than 11,000 square feet of space and can accommodate up to 1,200 people for banquets. The hotel also boasts a wood-paneled boardroom for guests' executive conferences. Audiovisual equipment is available, and, of course, the hotel staff includes experienced convention coordinators to attend to every meeting need.

Hotel services include 24-hour room service, concierge for all guests, and a corporate club suite. Sheraton Music City's full-service restaurant, the Belair Dining Room, features an award-winning Sunday brunch. Private dining rooms are available for special occasions. Hotel chefs are well known for their daily seafood specials.

The Sheraton Music City Hotel invites guests to discover why the hotel has earned the AAA Four-Diamond Award and the prestigious Gold Key Award of *Meetings and Convention* magazine for excellence in meeting services. The hotel's efforts are geared to ensure quality, consistency, and sincerity of the service staff. At Sheraton, little things mean a lot.

Bibliography

Art Work of Nashville 1894-1901. W.H. Parish Publishing Co. Chicago, IL, 1984.

Bartik, Tim. "Saturn and State Economic Development," Vanderbilt University. Nashville, TN, 1987.

Beckham, Joanne. "The Railroad Industry in Transition," *Advantage Magazine*. August 1989.

Birnbach, Lisa. *College Book*. Ballantine. New York, 1984.

Burch, David L. "The Truth About Startups," *INC*. January 1988.

Corlew, Folmsbee and Mitchell. *Tennessee, A Short History*. University of Tennessee Press. Knoxville, TN, 1969.

Dick, Margaret. "Miracle on Second Avenue," *Advantage Magazine*. November 1979.

Doyle, Don H. *Nashville In The New South*. University of Tennessee Press. Knoxville, TN, 1985.

--------. *Nashville Since the 1920s*. University of Tennessee Press. Knoxville, TN, 1985.

Egerton, John. *Nashville: The Faces of Two Centuries*. Plus Media Incorporated. Nashville, TN, 1979.

Longino, Marianne. "Nashville Report," *Pace Magazine*. 1986.

Lynch, Amy. "The Company Was Olympian," *Nashville Magazine*. November 1983.

--------. "Middle Tennessee Mecca," *Nashville Magazine*. March 1984.

Mikelbank, Peter. "Printing Industry Outlook," *Advantage Magazine*. June 1987.

Runyon, Marvin T. Speech to Foreign Correspondents Club of Japan. March 29, 1983.

Smith, Reid. *Majestic Middle Tennessee*. Paddle Wheel Publications. Prattville, AL, 1975.

Smith, Timothy K. "Nashville Is Booming and a Little Worried How It Will Turn Out," *Wall Street Journal*. January 7, 1987.

Sunshine, Linda, and John W. Wright. *The Best Hospitals in America*. Avon Books. New York, 1987.

"Top Twenty Employers," *Nashville Business Journal*. Nashville, TN, December 12-16, 1988.

Vise, Joyce. "Nashville Public Schools: How Good Are They?," *Nashville!*. April 1988.

Wissner, Sheila. "TSU Reaching, Researching Stars," *The Tennessean*. August 25, 1989.

Thanks to the following for the use of their files and publications.

Baptist Hospital
City of Nashville
Fisk University
Metro Public Schools
Nashville Area Chamber of Commerce
Tennessee State University
Urban Development Office of Metropolitan Development and Housing Agency
Vanderbilt University
Vanderbilt University Medical Center

Patrons

The following individuals, companies, and organizations have made a valuable commitment to the quality of this publication. Windsor Publications gratefully acknowledges their participation in *Nashville: Upbeat and Down to Business*.

American General Life and Accident Insurance Company*
The Bailey Company*
Barcus Nugent Consulting*
Bridgestone*
Corroon & Black Corporation*
Dearborn & Ewing*

DET Distributing Co.*
Edwards + Hotchkiss Architects, Inc.*
Gould Turner Group, P.C.*
Walter Knestrick Contractor, Inc.*
Kraft Bros., Esstman, Patton & Harrell*
Lee Company*
Loews Vanderbilt Plaza Hotel*
Maryland Farms*
Maxwell House Hotel*
William M. Mercer, Incorporated*
Metropolitan Nashville Airport Authority*
The Nashville Banner*

Nashville Electric Service*
Nissan Motor Manufacturing Corporation U.S.A.*
Robert Orr/SYSCO*
Reemay, Inc.*
Sheraton Music City Hotel*
The Tennessean*
Textron Aerostructures*
WLAC AM/FM*

*Participants in Part Two: *Nashville: Upbeat and Down to Business*. The stories of these companies and organizations appear in chapters 8 through 12, beginning on page 96.

Directory Of Corporate Sponsors

American General Life and Accident
Insurance Company, 126-127
American General Center
310 8th Street
Nashville, TN 37250
615/749-1000
Carroll D. Shanks

The Bailey Company, 113
501 Cowan Street
Post Office Box 80565
Nashville, TN 37208
615/242-0351
Gordon Morrow

Barcus Nugent Consulting, 122-123
404 James Robertson Parkway,
Suite 1500
Nashville, TN 37219
615/256-4100
Terry Nugent

Bridgestone, 110-111
Post Office Box 140991
100 Briley Corners
Nashville, TN 37214-0991
615/391-0088
Trevor C. Hoskins

Corroon & Black Corporation, 125
301 Plus Park Boulevard
Nashville, TN 37202
615/367-9702
Thomas P. Lawrence

Dearborn & Ewing, 128-129
One Commerce Place, Suite 1200
Nashville, TN 37239
615/259-3560
Joseph Barker

DET Distributing Co., 116
301 Great Circle Road
Nashville, TN 37228
615/244-4113
David Earls

Edwards + Hotchkiss Architects, Inc.,
136-137
The Horsebarn at Maryland Farms
Brentwood, TN 37027
615/377-3111
Richard Hotchkiss

Gould Turner Group, P.C., 138
1801 West End Avenue
Nashville, TN 37201
615/327-3122
Steve Turner

Walter Knestrick Contractor, Inc., 134
2617 Grand View Avenue
Nashville, TN 37222-1269
615/259-3755
Lee Munz

Kraft Bros., Esstman, Patton & Harrell,
120-121
404 James Robertson Parkway,
Suite 1200
Nashville, TN 37219
615/242-7351
H. Safer

Lee Company, 135
322 Wilhagen Road
Nashville, TN 37217
615/367-9206
Wallace Lee

Loews Vanderbilt Plaza Hotel, 144-145
2100 West End Avenue
Nashville, TN 37203
615/320-1700
Irwin E. Fisher

Maryland Farms, 132-133
109 Westpark Drive, Suite 250
Brentwood, TN 37027
615/373-2000
Nelda Bates

Maxwell House Hotel, 142-143
2025 Metro Center Boulevard
Nashville, TN 37228
615/259-4343
Karen Groom

William M. Mercer, Incorporated, 124
1300 Third National Financial Center
424 Church Street
Nashville, TN 37219
615/259-1300
David Hollis

Metropolitan Nashville Airport
Authority, 99
One Terminal Drive, Suite 501
Nashville, TN 37214
615/275-1608
Stephan S. Foust

The Nashville Banner, 102
1100 Broadway
Nashville, TN 37203
615/259-8800
Irby Simpkins, Jr.

Nashville Electric Service, 98
Church Street at 13th Avenue
Nashville, TN 37203
615/747-3513

Nissan Motor Manufacturing
Corporation U.S.A., 112
Nissan Drive
Smyrna, TN 37167
615/459-1400
Gail O. Neuman

Robert Orr/SYSCO, 108-109
One Hermitage Plaza
Post Office Box 305137
Nashville, TN 37230
615/350-7100
Howard Stone

Reemay, Inc., 114-115
70 Old Hickory Boulevard
Post Office Box 511
Old Hickory, TN 37138
615/847-7000

Sheraton Music City Hotel, 146-147
777 McGavock Pike
Nashville, TN 37214
615/885-2200
Henry Rather

The Tennessean, 101
1100 Broadway
Nashville, TN 37203
615/259-8800
Joe Pepe

Textron Aerostructures, 106-107
1431 Vultee Boulevard
Post Office Box 210
Nashville, TN 37217
615/361-2000
Clint Smith

WLAC AM/FM, 100
10 Music Circle East
Nashville, TN 37203
615/256-0555
Elizabeth Yoder

Index